Lecture Notes in Computer Science 584
Edited by G. Goos and J. Hartmanis

R. E. Zippel (Ed.)

Computer Algebra and Parallelism

Second International Workshop
Ithaca, USA, May 9-11, 1990
Proceedings

Springer-Verlag
Berlin Heidelberg New York
London Paris Tokyo
Hong Kong Barcelona
Budapest

Series Editors

Gerhard Goos
Universität Karlsruhe
Postfach 69 80
Vincenz-Priessnitz-Straße 1
W-7500 Karlsruhe, FRG

Juris Hartmanis
Cornell University
Department of Computer Science
5149 Upson Hall
Ithaca, NY 14853, USA

Volume Editor

Richard E. Zippel
Cornell University, Dept. of Computer Science
5138 Upson Hall, Ithaca, NY 14853, USA

CR Subject Classification (1991): I.1, C.4

ISBN 3-540-55328-2 Springer-Verlag Berlin Heidelberg New York
ISBN 0-387-55328-2 Springer-Verlag New York Berlin Heidelberg

Typesetting: Camera ready by author
Printing and binding: Druckhaus Beltz, Hemsbach/Bergstr.
45/3140-543210 - Printed on acid-free paper

Preface

From May 9 to May 11, 1990, the Army Research Office sponsored Mathematical Sciences Institute at Cornell University hosted a workshop on *Computer Algebra and Parallelism* (CAP '90), which immediated followed a companion workshop on Computer Algebra and Differential Equations. The first meeting of these two workshops was sponsored by the University of Grenoble, in May and June 1988, and was conceived by the 1990 co-organizers, J. Della Dora and E. Tournier (Grenoble) and M. Singer (North Carolina State). This year's organizing committee also included J. Fitch (Bath, England), E. Kaltofen (Rensselaer Polytechnic Institute), and R. Zippel (Cornell).

The use of parallel computers in symbolic mathematical computation (computer algebra) lags somewhat behind their use in scientific computation (numerical mathematics). The main goal of the CAP workshops is to address the significance of parallel computing in the discipline of symbolic mathematical computation. This workshop pointed to several pioneering efforts to design new parallel algebraic algorithms from a theoretical point of view, to design parallel programming environments suited for symbolic computation, and to attack large symbolic problems by using parallel computers such as the Cray Y-MP or the Connection Machine.

Eight of the presentations given at the workshop are represented in these proceedings. These papers are divided into three groups: systems designed to support parallel symbolic computation, parallel algorithms for dealing with finite fields and algebraic numbers, and parallel implementations of Gröbner basis algorithms.

Three papers constitute the first section. The first paper, by Küchlin, discusses S-threads, a mechanism that combines a control thread with a heap fragment. This extension of the familiar "thread" concept simplifies parallel symbolic computing significantly. Seitz discusses a different mechanism that is based on the "futures" concept popularized by Multilisp. Seitz has extended ALDES to incorporate futures in a distributed network of workstations. The final paper in this section, by Roch, discusses a symbolic computing environment specially designed for parallel systems. This system was implemented on two different tightly coupled parallel systems, an FPS hypercube and a TELMAT Meganode. Roch also discusses his experiences with these systems.

The first paper in the second section discusses a unique application of the Connection Machine to perform computations in finite fields of the form $GF(p^m)$. Sibert, Mattson and Jackson use a table driven discrete logarithm technique to perform the arithmetic computations. They then use their tools to count the number of zeroes of large families of polynomials. Weeks discusses a fascinating approach to parallel computation with algebraic numbers. The usual way to represent an algebraic number is by presenting its minimal polynomial, i.e., a vector of the symmetric functions of its conjugates. Weeks observed that by using the sums of powers of the conjugates one can perform many operations very quickly and with a great deal of parallelism.

In Chapter 6, Collins, Johnson and Küchlin present an implementation of a parallel real root isolation technique that uses interval bisection and Descartes' rule signs. This algorithm is implemented on an Encore Multimax using the PARSAC technology developed by Küchlin and discussed in Chapter 1.

There are two papers in the final section on Gröbner bases. The first, by Neun and Melenk, discusses an implementation of the Gröbner basis algorithm on the Cray supercomputer. Neun and Melenk observed that a substantial amount of time is spent performing arithmetic with very large numbers. They have developed effective mechanisms for taking advantage of the vector units of the Cray both for arbitrary precision arithmetic computations and in the garbage collector. The final paper, by Sénéchaud examines the effect of two different multiprocessor topologies (a ring and a hypercube) on parallel Gröbner basis algorithms.

In a concluding discussion among the participants, failures and successes in bringing parallel computing into computer algebra were considered. The computer algebra community is widely scattered throughout the United States and the rest of the world. For three days, this workshop concentrated researchers active in the parallel techniques for symbolic computation into one place. It is anticipated that the twin workshops will be held again in 1992 in Europe, at which time the effect of this gathering will certainly be apparent.

January 1992

Erich Kaltofen
Chairman

Richard Zippel
Editor

Contents

1

The S-Threads Environment for Parallel Symbolic Computation

Wolfgang Küchlin[1]

1.1 Introduction

1.1.1 Overview

This paper presents a programming environment, based on *threads of control*, that is suitable for parallel *symbolic* computation on shared memory multiprocessors. The S-threads system offers a solution to the problem of whether to have heap memory shared and global, or distributed, and local to threads. The memory structure makes it particularly easy to reclaim, without garbage collection, all intermediate list memory used by an algorithm; under some additional restrictions, S-threads may also perform independent garbage collections. For empirical results motivating this work see [SLA89, Küc90].

The S-threads environment is being used in the construction of the PARSAC system [Küc90], a parallel version of the SAC-2 Computer Algebra System[2] [CL]. To date, in Summer 1990, PARSAC-2.1 contains parallel algorithms for integer multiplication [Küc90] and for isolating the real roots of integer polynomials [CJK90]; work on parallel multivariate polynomial g.c.d. calculation and on parallel root isolation of algebraic polynomials is under way. S-threads and PARSAC-2 are implemented on an Encore Multimax, based on the C Threads [CD87] environment emulated by Encore Parallel Threads [Enc88].

1.1.2 Motivation

Parallelizing symbolic algorithms is promising because many manipulations in Logic and Algebra are inherently concurrent, and at the same time they do not benefit from other hardware support such as floating point coprocessors. However, symbolic computation presents special problems for parallelization, such as extreme data dependency with unpredictable task lengths and memory demands, and the presence of global shared list memory with equally unpredictable demand for garbage collec-

[1]Department of Computer and Information Science, The Ohio State University, Columbus, OH 43210-1277, Email: Kuechlin@cis.ohio-state.edu

[2]For this purpose, SAC-2 was translated from ALDES [Loo76] to C.

tions. The S-threads environment attempts to meet these challenges by providing a portable, highly efficient parallelization environment for shared memory multiprocessors. While our implementation was done on an Encore Multimax minicomputer, workstations and servers with the same architecture are already on the market.

Shared memory multiprocessors are not as scalable as distributed memory machines. Nevertheless, they constitute a promising platform for parallel symbolic computation since they allow for fast distribution of tasks and data to physical processors. Therefore, task start-up times are low, so that the parallel grain size is smaller and more tasks can profitably be executed in parallel. This decreases the need for decisions whether to fork tasks in parallel, and increases the degree of exploitable parallelism. Further, fast dynamic load balancing by the operating system is feasible, so that the application programmer need not be concerned with task placement. Predictions about task size and actual control flow are all highly error-prone in symbolic computation. Finally, the heap can be allocated in shared memory, and list parameters can be passed to tasks by reference.

Task start-up times some order of magnitude below those of conventional processes have been achieved with *threads* (lightweight processes), as realized e. g. under the Mach operating system [ABB+86]. C Threads [CD87] are a minimal, though extendable, threads abstraction which is accessible from C as a system of library calls. The C Threads package provides primitives for forking and joining of threads, and for synchronization through mutex and condition variables. Portability can easily be achieved through Mach; a threads library can also be implemented directly with reasonable effort (cf. [Sam89]). An S-thread may be thought of as a C thread extended by a local environment containing a heap fragment and additional information.

This document describes Release 1 of the S-threads system as used in PARSAC-2.1. In Section 1.2, we introduce the system by the simple example of a parallel list sorting algorithm. In Section 1.3, we then give an overview of the design of S-threads and the current implementation. In Sections 1.4 and 1.5 we describe the functionality of the primitives for threads and memory manipulation, respectively; Section 1.6 gives further examples of their use.

Note

In the time between the original writing of this article and its publication, S-threads was ported to the Mach operating system, and several more SAC-2 algorithms were parallelized. Parallel multivariate polynomial g.c.d. calculation by the Brown-Collins method, including parallel Chinese Remaindering in $Z[x_1, \ldots, x_n]$ and parallel interpolation in $Z_p[x_1, \ldots, x_n]$, is treated in [Küc91a, Küc91b]. Parallel integer multiplication by the Karatsuba and 3-primes FFT methods is reported in [KLN91].

For each algorithm, the parallel speed-up was measured on a wide range of inputs. Many of these calculations posed heavy demands on the S-threads environment, using hundreds of threads and causing parallel garbage collections. Our experiments so far confirm that S-threads is a very suitable environment for parallel algebraic computation. In particular, its overhead is low enough to allow easy and efficient parallelization, and *preventive garbage collection* (see Section 1.3.2 below) is so effective and efficient that it appears attractive even for sequential systems.

Our experience with these large applications has led to some revisions of S-threads for Release 2. Most importantly, two new fork constructs have been added: a *pseudo-fork* now supports preventive garbage collection in sequential code, and a *task-fork*, already suggested in Section 1.3.2 below, facilitates concurrent conventional garbage collection. Our hierarchy of three fork constructs will now also enable a future extension of S-threads to cover hierarchical parallel machines such as networks of multiprocessor workstations. S-threads Release 2 and its parallel garbage collection methods are presented in [KN91]; a summary is contained in [Küc91b].

1.2 A Programming Example

We give an informal introduction to parallel programming in PARSAC using a simple example of list sorting. We also demonstrate in principle the conversion process from an existing sequential SAC-2 algorithm to an *efficient* parallel version.

Parallelism is introduced under S-threads by *forking* a new parallel thread of control from the executing thread with a call to `sthread_fork()`, which returns the ID of the child. The result of the child can be collected by *joining* it with `sthread_join()`, which may have to wait for the child to finish. Since only one pointer-sized argument can be passed to a function forked in parallel, a *thread shell*, which does the unpacking, is usually created for each parallel routine. Fork/join primitives alone provide for a form of parallel functional programming; we do not demonstrate other means of synchronization here. Precise explanations of all S-thread calls are given in Sections 1.4 and 1.5.

Each S-thread maintains its own heap fragment on which it allocates list cells. Function `space_reset()` resets the heap of the executing thread, freeing all cells. The memory organization which allows this technique of *preventive garbage collection* is explained in Sections 1.3 and 1.5.

For simplicity, our example was chosen at the SAC-2 list-processing level. Although the example demonstrates satisfactory speedup, parallelization of SAC-2 would normally be done on more complex algebraic functions. The SAC-2 list handling primitives are `COMP`, `FIRST`, and `RED`, corresponding to `cons`, `car`, and `cdr`, in LISP. `BETA` is the base for SAC-2 integers; a β-digit[3] is an integer i with $|i| < \beta$, i.e. i is a digit in base β notation. No further knowledge of SAC-2 data-structures is needed here.

---

[3] Sometimes (somewhat confusingly) also called β-integer.

```
#include <stdlib.h>    /* For random numbers with rand(). */
#include "sthreads.h"
#include "sac2init.h"  /* Defines SAC-2 globals. */

                         main()
{       sthread_init(main1);
}

                void     main1()
{int j=100, L, Lp, Ls, Lsp;
LOOP: /*(1) Generate a random list with a copy. */
        L=LBDrand(j);  Lp=flcopy(L);
/*(2) Sort L by parallel bubble-merge-sort.*/
        Ls=p_LBIbms(j,L);
/*(3) Sort Lp by sequential bubble-merge-sort.*/
        Lsp=LBIBMS(Lp);
/*(4) Clear memory. */
        space_reset();
/*(5) Iterate? */
        j += 100;  if (j <= 5000) goto LOOP;
}

                int      LBDrand(l)
int l;
/* List of beta-digits, random.
 * A list of l random beta-digits is generated.
 */
{int L,i,d;
/*(1) Initialize. */
        L = NIL;
/*(2) Generate list. */
        for (i=0; i<l; i++)
         {d=rand() % BETA; L = COMP(d,L);}
/*(3) Finish up. */
        return L;
}

                int      p_LBIbms(n,L)
int n,L;
/* List of Beta-Integers, parallel bubble-merge-sort.
 * L is a list of Beta-Integers, n=LENGTH(L).
 * Then L is destructively sorted into non-descending order.
 */
{int m, t, i, LP, L2, L1;
 int arg[2];
```

```
sthread_t th;
/*(1) Base of recursion: short list. */
        t=10;
        if (n<t) {LBIBS(L); return (L);}
/*(2) Split list. */
        LP = L; m = n/2;
        for (i=0; i< m -1; i++) LP=RED(LP);
        L2=RED(LP); RED(LP)=NIL;
/*(3) Recursion: sort parts. */
        arg[0]=n-m; arg[1]=L2;
        th=sthread_fork(p_LBIbms_shell, arg, NULL);
        L1=p_LBIbms(m,L);
        L2=sthread_join(th);
/*(4) Combine: merge sorted parts. */
        return LBIM(L1,L2);
}

                    int       p_LBIbms_shell(arg)
int arg[2];
{         return p_LBIbms(arg[0],arg[1]);
}
```

The SAC-2 function **LBIBMS** sorts a list of β-digits into non-decreasing order using a combination of merge-sort and bubble-sort. Auxiliary functions are **LBIBS** (List of Beta-Integers, Bubble-Sort) and **LBIM** (List of Beta-Integers, Merge). **LBIBMS** cuts an input list **L** of length $n > 10$ into m pieces of length less than 10, then sorts each piece with **LBIBS**, and finally merges the sorted pieces to the final result using **LBIM**.

Straight-forward parallelization might apply **LBIBS** in parallel m times, and also apply **LBIM** in parallel during each of the $\log n$ phases of merging. Each of the parallel processes would form a separate S-thread; their ID's could be kept in lists (using the primitives described in Section 1.4.11) and then joined one after another.

A recursive approach is far simpler to program: **p_LBIbms** sorts the input list **L** by **LBIBS** if **L** has length less than 10; if it is longer, **L** is split into **L1** and **L2** which are sorted recursively in parallel, after which the results are merged. This obviates the need for the two dynamic lists of thread ID's because the static description of the program has to fork and join only one S-thread. Note that forking C threads would even be sufficient because **p_LBIbms** does not allocate any list cells.

The run times[4] of **p_LBIbms** *vs.* **LBIBMS** are shown in Figure 1.1. So far our effort has yielded a slow-down rather than a speed-up, because we ignored the parallel grain size afforded by our implementation. With a combined fork/join cost between 2 and 3*ms* (see [Küc90]), we can limit the parallel overhead to below 10% if we

---

[4] All runs used maximally 8 of 12 NS32332 processors of an Encore Multimax. All times were taken from the microsecond wall clock timer and rounded to milliseconds.

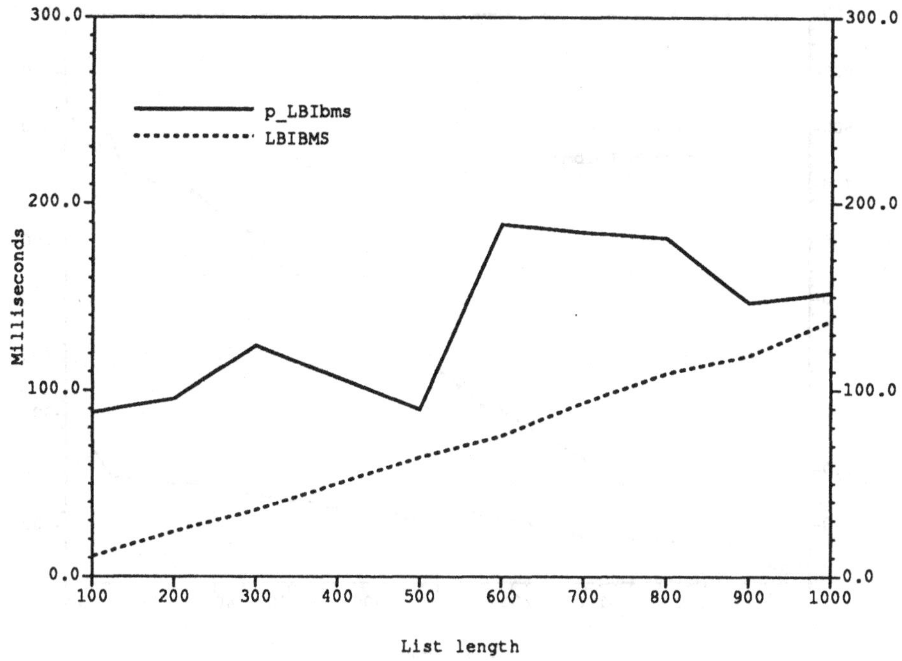

Figure 1.1. p_LBIbms *vs.* LBIBMS

choose a grain size of $30ms$. Figure 1.1 shows that **LBIBMS** is above the grain size if input lists are longer than about 300 cells. This determines our *parallel cut-off point* which is in addition to the sequential cut-off point of 10 cells between **LBIBMS** and **LBIBS**. We therefore replace step (3) of **p_LBIbms** by the following.

```
/*(3) Recursion: sort parts. */
/*(3.1) Below parallel grain size? */
        if (m<300)
         {L2=p_LBIbms(n-m,L2); L1=p_LBIbms(m,L);}

/*(3.2) Above parallel grain size? */
        else
        {arg[0]=n-m; arg[1]=L2;
         th=sthread_fork(p_LBIbms_shell, arg, NULL);
         L1=p_LBIbms(m,L);
         L2=sthread_join(th);
        }
```

The new runtimes are shown in Figure 1.2 for inputs of up to 5000 list cells. The following table samples run-times, speed-up, and efficiency, at critical points where the number of forked threads changes.

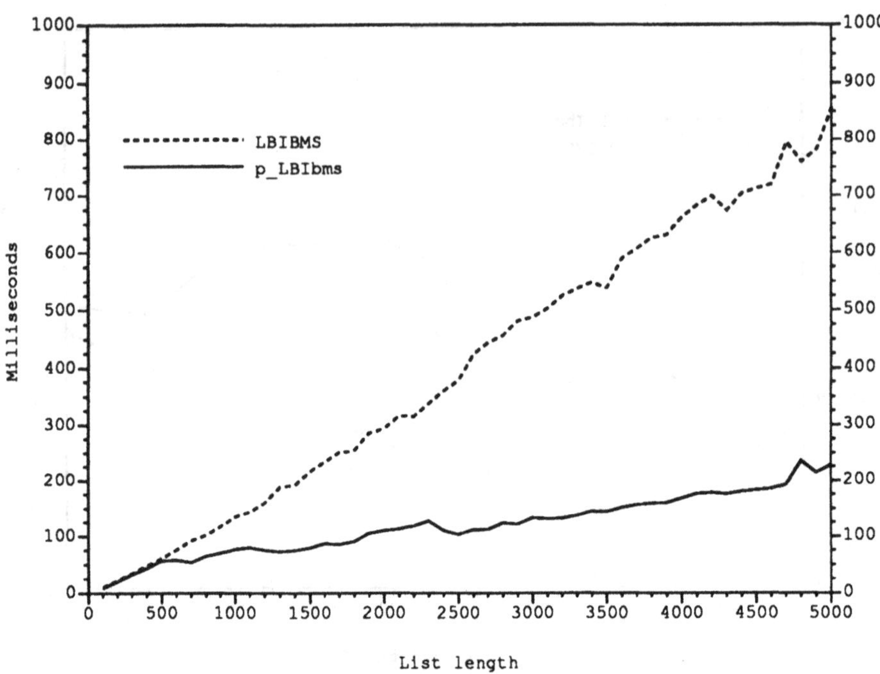

Figure 1.2. Optimized `p_LBIbms` *vs.* `LBIBMS`

#cells	LBIBMS	p_LBIbms	speed-up	efficiency	#processors
500	60*ms*	57*ms*	1.05	52.5%	2
600	76*ms*	58*ms*	1.31	32.8%	4
1100	143*ms*	80*ms*	1.79	44.7%	4
1200	159*ms*	75*ms*	2.12	26.5%	8
2300	337*ms*	127*ms*	2.65	33.2%	8
2400	359*ms*	110*ms*	3.26	40.8%	8
4700	795*ms*	192*ms*	4.14	51.7%	8
4800	760*ms*	235*ms*	3.23	40.4%	8

1.3 S-threads System Design

The S-threads system contains a *threads subsystem* providing threads of control, and a *space subsystem* providing list memory handling. The specification has been left incomplete on purpose: Synchronization primitives are not covered and list handling only specifies a list cell allocation function. The general assumption is, that the threads subsystem will be implemented on top of some parallel computation substrate that also provides synchronization, and that an existing sequential list processing system will be imported into the space subsystem to support a particular

1.3.1 The PARSAC Implementation

Our current implementation of S-threads builds respectively on C Threads and on the list processing (LP) package of SAC-2 (whose list memory is known as *SPACE*). Little knowledge about SAC-2 is needed here, but we do assume familiarity with C Threads in the following. Empirical results have been reported in [Küc90].

From a SAC-2 point of view, S-threads are a new system foundation at, and below, the SAC-2 List-Processing level. S-threads replace algorithms **COMP** (cell allocation), **BEGIN1** (LP initialization), and **GC** and **MARK** (garbage collection). All other SAC-2 functions at the LP level or above can be executed unmodified by a thread of control.

From a C Threads point of view, S-threads provide an enhancement for list processing. Every S-thread is an extension grafted onto a substrate C thread, which makes the S-thread about twice as "heavy" (i.e. slower to fork and join) as the substrate. S-threads are thus largely compatible with C Threads. Pure C threads can be used alongside with S-threads to perform computations which do not allocate list cells. (However, it is an error to use any C Thread call, as in section 4 of [CD87], on the substrate thread of an S-thread.) Any system calls described in [CD87] other than the thread handling primitives can be used with S-threads; in particular, S-threads can be synchronized with the mutex and condition handling primitives of the C Threads environment.

Release 1 of S-threads, and hence PARSAC-2.1, do not yet provide garbage collection in the traditional sense. With garbage collection in place, the S-threads environment can be used with little knowledge of its memory organization. However, Section 1.5 describes primitives of the parallel space subsystem which can be used to obtain a weaker form of garbage collection. This form, called ***preventive garbage collection***, was sufficient for several non-trivial experiments reported in [Küc90, CJK90], as well as others. Below, we describe the memory organization underlying these primitives.

1.3.2 Memory Organization

When combining C Threads with list-processing, it must be decided whether to keep list memory global shared, with a single available cell list, or divided into thread-local partitions.

Keeping the heap shared has the advantage that it allows passing list structures by reference. (Lists may be very large, overlapping, or cyclic, and the recipient may only want to read them.) It has the serious[5] disadvantage that list cell allocation easily becomes a sequential bottleneck if there is only one shared resource of cells, a global available cell list. In addition, all threads need to be collected at a barrier if garbage collection is necessary. This is another formidable problem because it is impossible to predict in general how many threads are active at any given time. It

[5] For PARSAC-2 applications this effect is indeed prohibitive [Küc90].

if garbage collection is necessary. This is another formidable problem because it is impossible to predict in general how many threads are active at any given time. It is yet another problem to orchestrate parallel garbage collection even if all threads can be suspended at a barrier first.

Separating the list memory into thread-local partitions has the advantage that thread-local garbage collection is now possible. It has the disadvantage that list parameters need to be copied, including relocation of pointers, when they are passed between threads. It also does not make economical use of the heap as a scarce resource, because memory demand may not be uniform and only a fraction of threads may be executing at any time.

Under S-threads, global heap memory is divided into *pages*, which are then dynamically allocated by threads as local page-sets. We thus have a global page-pool with a free list of pages, and each S-thread owns a local page-set, constituting its private list memory, with a local unlocked available cell list. All page transfers are of course by reference only. Allocating cells by pages alleviates the bottleneck problem through decreased demand for the allocation lock, and by amortizing page allocation cost over many list cells.

The page-set of an S-thread can be garbage collected independently in a straightforward way if there are no external references into the set, and no references out of the set. This can be achieved by adopting a functional programming style, and by copying input parameters. More complicated garbage collection, which is necessary if there are references crossing page-sets, is beyond the scope of this paper.

The need for traditional garbage collections can be decreased substantially by a weaker scheme based on the observation that, at the end of an algorithm, its page-set contains all intermediate data-structures as well as the result. If the result is copied onto new pages, the old page-set contains only garbage and can be returned to the page-pool. Given the usual intermediate expression swell of symbolic computation, this page collection scheme is very economical. In effect, compacting garbage collection is carried out at carefully selected times, when little is to be marked, determined by static knowledge about the program. This is in contrast to usual garbage collection which occurs at random times and therefore has to compute similar knowledge dynamically, at a high price, by the marking algorithm. Due to the similarity with preventive maintenance, including the cost benefits, we speak of *preventive garbage collection*.

In summary, we have the following basic mechanisms for passing a list structure L, stored on page-set P, from thread A to thread B:

1. *By reference:* a reference to L is passed.
2. *By transfer:* (a reference to L is passed and) P is transferred (by reference) from A to B.
3. *By copy and transfer:* L is first copied to L' on page-set P'; then L' is passed by transfer.

These three mechanisms for passing list parameters give rise to the following four important modes for parallel procedure calls in PARSAC.

1. *Reference in—transfer out:* List input parameters are passed by reference from parent to child, while list output parameters are transferred back on the entire page-set M owned by the child upon termination.

2. *Reference in—copy transfer out:* Input parameters are passed by reference; output parameters are passed by copy transfer.

3. *Copy transfer in—transfer out:* List input parameters are passed by copy transfer; output parameters are passed by transfer.

4. *Copy transfer in—copy transfer out:* Both input and output list parameters are passed by copy transfer.

The critical issue in choosing a mode is that of list *lifetime*: if a thread creates a list cell whose lifetime is longer than that of the thread, then the page containing the cell must be transferred to a thread with a longer lifetime (e. g. the parent). In functional programs, this affects only the output parameters. If a thread updates a global data-structure such as a symbol table, its entire page-set may have to be saved in order to protect the cells with longer lifetime.

Modes 1 and 2 are supported and used in PARSAC-2.1; modes 3 and 4 will be supported Release 2. Modes 1 and 3 extend the lifetime of all list cells beyond that of the child, and therefore enable global list updates. Modes 2 and 4 perform preventive garbage collection of the child.

Functions called in modes 1 and 2 are threads operating in shared memory. Functions called in modes 3 and 4 effectively form *tasks* with private memory. It follows that their memory segment can be garbage collected independently. They can also be executed on a separate processor in a distributed environment if page transfer is implemented by message passing.

Section 1.6 contains three small examples on the use of preventive garbage collection and calling modes in PARSAC-2.1.

1.4 Threads of Control for Symbolic Computation

For the S-threads interface specification we proceed in close analogy to sections 3 and 4 of [CD87]. Every **sthread**... call is the logical equivalent of the corresponding **cthread**... call, except for differences stated explicitly as S-threads specific.

```
typedef ... sthread_t;
```

The **sthread_t** type is a pointer that uniquely identifies an S-thread and may serve as its ID (out of a sparse name-space). Actually, it is a "handle" of the respective *S-thread control block*, a data-structure that contains the private environment of each S-thread.

S-thread primitives may not be assumed to be synchronized; it is an error to call two primitives concurrently on the same S-thread. It is wise for parents to watch over their children and not let other threads take control over them.

1.4.1 INITIALIZATION

1.4.1.1 STHREADS.H

```
#include "sthreads.h"
```

The header file **sthreads.h** specifies the syntax of the S-threads interface. Inclusion of **<cthreads.h>** is implicit.

1.4.1.2 STHREAD_INIT

```
void sthread_init(func)
void (*func) ();
```

This procedure initializes the S-threads system.

S-threads specific: It starts a thread and transfers control to function **func()** which executes on that thread. Typically, the main program **main()** contains just one line **sthread_init(main1)**, where **main1** is (a pointer to) a procedure that takes the role of **main()** in a conventional C program.

Remark: The initialization of C Threads is done by **cthread_init()** without any arguments, and control stays in the calling procedure, typically **main()**. Under Mach every UNIX process is executed by a thread within a Mach task; therefore, **main()** itself is already executed by a thread. Where threads are provided as an application package, **main()** is executed by a conventional UNIX process. Initialization of C threads then carves the threads system out of that process, and only **cthread_init(func)** is able to start ***func** on the first thread.

1.4.2 STHREAD_FORK

```
sthread_t sthread_fork(func, arg, copyfunc);
int (*func) ();
int *arg;
int (*copyfunc) ();
```

Creates a new thread of control for list-processing functions, executing **(*func) (arg)** concurrently with the caller. Arguments to **(*func)** must be "packaged" in the parent thread (by the programmer) and passed through a package handle. (It is an error if the lifetime of the package structure can be shorter than the lifetime of the forked child.) The call returns the S-thread handle of the forked thread. Every S-thread must be joined or detached exactly once. The result of the forked S-thread is determined by **sthread_exit()** (see below). An **sthread_exit(v)** is implicit after **v=(*func) (arg)** has been computed on the forked child.

S-thread specific: The *virtual result* of the forked thread is the parameter of the call to **sthread_exit()** executed by the child. The *actual result* is computed from the virtual result **v** by **sthread_exit(v)**, depending on the passing mode. The passing mode is determined as follows, with the help of the (output) copy function **copyfunc**:

1. If **v** is simple (an atom or an empty list), the mode is "pass by value."

2. If **v** is compound (a non-empty list), and **copyfunc==NULL**, the mode is "pass by transfer."

3. If **v** is compound and **copyfunc != NULL**, the mode is "pass by copy transfer."

Remark: For PARSAC use, it is sufficient if parameters are of type integer, pointer to integer, or function yielding integers; the C Threads defined **any_t** could have been used for more generality. The rationale behind copy functions is given in Section 1.3.2.

1.4.3 STHREAD_EXIT

```
void sthread_exit(v);
int v;
```

Causes termination of the executing S-thread with result **v**.

S-thread specific: Let M be the page-set of the S-thread and **copyfunc** its copy function. The virtual result **v** is converted to the actual result **a** of the S-thread as follows:

1. If the mode is "pass by value," then **a=v**, and M is freed.

2. If the mode is pass by transfer, then **a=v**, and M is transferred to the joining S-thread.

3. If the mode is pass by copy and transfer, then **a=(*copyfunc) (v)** is computed onto new memory pages which are transferred to the joining S-thread, and M is freed.

1.4.4 STHREAD_JOIN

```
int sthread_join(st)
sthread_t st;
```

sthread_join(st) suspends until the S-thread **st** exits; it then returns the actual result R of **st**.

S-thread specific: If R is compound, the caller's local list memory is augmented by the pages on which R exists; the caller's available cell list also inherits all available cells on those pages.

1.4.5 STHREAD_DETACH

```
void sthread_detach(st)
sthread_t st;
```

The parent's ties to S-thread **st** are severed, i.e. **st** will never be joined and its result, if any, will be discarded upon exit.

Remark: Not yet implemented.

1.4.6 STHREAD_YIELD

```
void sthread_yield()
```

Hint to the scheduler to switch threads in a coroutine emulation of the system.

1.4.7 STHREAD_SELF

```
sthread_t sthread_self()
```

Yields the handle (ID) of the executing S-thread.
Remark: See also **sthread_index()** below.

1.4.8 STHREAD_SET_DATA, STHREAD_DATA

```
void sthread_set_data(st,d);
sthread_t st;
int d;
int sthread_data(st);
sthread_t st;
```

Set/get user data of S-thread. These primitives allow the user to store data associated with the S-thread **st**. Lifetime of the store is as long as that of the S-thread.

1.4.9 STHREAD_INDEX

```
int sthread_index(st);
sthread_t st;
```

S-thread specific: Get S-thread index. The index is taken from the dense namespace 0...n and may serve as an alternative ID suitable for debugging purposes or as an index into an array.

1.4.10 STHREAD_KILL

```
void sthread_kill(st);
sthread_t st;
```

S-thread specific: The S-thread **st** is given the kiss-of-death. It will be forcefully terminated the next time it issues a request for a new memory page, and in any case it will exit with result 0. It is illegal to kill any S-threads that one has not forked, and it is illegal to kill any child that has non-detached children. {**sthread_kill(st); sthread_detach(st);**} gets rid of S-thread **st** entirely.

Remark: This primitive has been kept restrictive in order to keep its implementation simple. It is useful for aborting parallel processes after a dynamically computed event has occurred, such as in parallel search, or in computing by multiple homomorphic images. Note that there is no corresponding C Thread primitive.

1.4.11 STHREAD_PUSH, STHREAD_POP, STHREAD_APPEND

```
void sthread_push(st, Q);
sthread_t st, Q;
sthread_t sthread_pop(Q);
sthread_t Q;
void sthread_append(st, Q);
sthread_t st, Q;
```

S-thread specific: S-threads may be organized into LIFO and FIFO lists during their lifetime. Only one list may be maintained at any one time, but all functions can be used on the same list. After an **sthread_push(st, Q)**, thread **st** is head of a list with tail **Q**; **sthread_pop(Q)** yields the remaining list after head element **Q** has been removed; **sthread_append(st, Q)** appends S-thread **st** at the end of list **Q**.

Remark: S-thread lists are not SAC-2 data-structures. These primitives can be used for barrier synchronization where an unpredictable number of children is first forked and then joined one after the other. Lists are disrupted if any member is joined.

1.5 The List Space

Functions in this subsystem either manipulate (SAC-2) lists or sets of pages of list cells. Only the former are really needed for programming with S-threads; the latter enable the sophisticated user to implement preventive garbage collection outside of S-thread forks.

In addition to the list handling primitives described below, all SAC-2 list functions can be used, with the notable exception of MARK() and GC().

In the following, *list* always denotes a SAC-2 list. A *SAC-2 object* is a SAC-2 atom, or a list of SAC-2 objects. All SAC-2 objects or their handles are C integers; in particular, a C float variable or a C pointer are *not* SAC-2 objects. It is illegal to construct lists containing non-SAC-2 objects (because SAC-2 list manipulation functions may get confused).

1.5.1 COMP

```
int COMP(a,L)
int a, L;
```

List composition. a is a SAC-2 object, L is a list. Then the result is the list with a as its head and L as its tail.

S-thread specific: The new list cell is allocated from the available cell list A in the local memory of the executing S-thread S. If A is empty, S's local memory is either increased by one page from the global page list, or else garbage collection occurs.

Remark: Note that several threads can simultaneously allocate list cells through

COMP. The garbage collection alternative is not implemented in Release 1 of S-threads. We expect to experiment with various strategies for the decision between paging and collecting.

1.5.2 LCOPY

```
int lcopy(L)
int L;
```

S-thread specific: List copy. `L` is a SAC-2 list structure in the form of a tree. Then the result of the function is a complete copy of `L`.

Remark: This is a popular parameter to `sthread_fork()`, suitable for most cases. If `L` is a dag rather than a tree, overlapping parts will be replicated several times. It is illegal to use `lcopy()` on cyclic lists. Beware of hidden cycles through property lists in the SAC-2 symbol system.

1.5.3 FLCOPY

```
int flcopy(L)
int L;
```

S-thread specific: Flat list copy. `L` is a linear SAC-2 list of atoms. Then the result of the function is a complete copy of `L`.

Remark: More efficient than `lcopy()` for linear lists such as long integers. More efficient copy functions for SAC-2 data-types such as polynomials will be included in later releases. The user may supply special copy functions for special data-structures.

1.5.4 SPACE_DETACH

```
int space_detach()
```

S-thread specific: Detach space from S-thread. The space pages of the executing S-thread are detached and returned as result.

Remark: A collection of pages is a SAC-2 object; more precisely, it is a linear list with undefined elements.

1.5.5 SPACE_ATTACH

```
void space_attach(M)
int M;
```

S-thread specific: Attach page-list to S-thread. The list `M` of memory pages is included in the local memory of the executing S-thread.

1.5.6 SPACE_FREE

```
void space_free(M)
int M;
```

S-thread specific: Free space. The list `M` of space pages is returned to the list of free pages on the global heap.

1.5.7 SPACE_RESET

```
void space_reset()
```

S-thread specific: Reset local space. The local memory pages of the executing thread are returned to the list of free pages on the global heap.

Remark: Shorthand for `space_free(space_detach())`.

1.6 Examples

1.6.1 Preventive Garbage Collection

The following fragment of (sequential) code performs preventive compacting garbage collection on intermediate memory used by the SAC-2 function IPRODK.

```
M0=space_detach();                          /* Start with clean slate.*/
K=IPRODK(I,J);                          /* K=I*J by Karatsuba's method.*/
M1=space_detach();                    /* All lists allocated by IPRODK.*/
K1=flcopy(K);                        /* Copy K onto fresh memory pages.*/
space_free(M1);                        /* Now M1 contains only garbage.*/
space_attach(M0);                  /*Bring M0 back into the environment.*/
```

1.6.2 Parallel Calls

The following code fragments illustrate the two parallel calling modes of PARSAC-2.1. Since IPRODK takes more than one input parameter, the unpacking is done in a separate *thread-shell.*

```
int p_Iprodk_shell(arg);
int arg[2];
{return IPRODK(arg[0],arg[1]);}
```

1.6.2.1 Reference In—Transfer Out

```
sthread_t st;
int arg[2], I, J, K;
...
arg[0]=I; arg[1]=J;
st=sthread_fork(p_Iprodk_shell,arg,NULL);
...
K=sthread_join(st);
```

1.6.2.2 Reference In—Copy Transfer Out

```
sthread_t st;
int arg[2], I, J, K;
...
arg[0]=I; arg[1]=J;
st=sthread_fork(p_Iprodk_shell,arg,flcopy);
...
K=sthread_join(st);
```

Acknowledgements

I am indebted to the Computer Algebra group at Ohio State, especially to Prof. George Collins and to Jeremy Johnson, for many fruitful discussions. The implementation would have been impossible without the ALDES-to-C translator written by Hoon Hong. I also wish to thank my student Mike Ulm for his help and criticism as a first user of the S-threads environment.

This material is based upon work supported by the Ohio State University Office of Research and Graduate Studies under Award No. 221152, and by the National Science Foundation under Award No. CCR-9009396.

1.7 References

[ABB+86] Mike Accetta, Robert Baron, William Bolosky, David Golub, Richard Rashid, Avadis Tevanian, and Michael Young. Mach: A new kernel foundation for UNIX development. In *Proc. Summer USENIX Conference*, July 1986.

[CD87] Eric C. Cooper and Richard P. Draves. C threads. Technical report, Computer Science Department, Carnegie Mellon University, Pittsburgh, PA 15213, July 1987.

[CJK90] George E. Collins, Jeremy Johnson, and Wolfgang Küchlin. Parallel real root isolation using the coefficient sign variation method. *These proceedings.*

[CL] G. E. Collins and R. G. K. Loos. SAC-2 system documentation. On-line documentation and program documentation. In Europe available from: Prof. R. Loos, Universität Tübingen, Informatik, D-7400 Tübingen, W-Germany. In the U.S.A. available from: Prof. G. E. Collins, Ohio State University, Computer Science, Columbus, OH 43210.

[Enc88] Encore Computer Corp. *Encore Parallel Threads Manual*, January 1988.

[KLN91] Wolfgang W. Küchlin, David Lutz, and Nicholas J. Nevin. Integer multiplication in PARSAC-2 on stock microprocessors. In *AAECC-9: Ninth Int. Symp. on Applied Algebra, Algebraic Algorithms, and Error-Correcting Codes*, LNCS, New Orleans, LA, October 1991. Springer-Verlag. (To appear).

[KN91] Wolfgang W. Küchlin and Nicholas J. Nevin. On multi-threaded list-processing and garbage collection. Technical Report OSU-CISRC-3/91-TR11, Computer and Information Science Research Center, The Ohio State University, Columbus, OH 43210-1277, March 1991.

[Küc90] Wolfgang W. Küchlin. PARSAC-2: A parallel SAC-2 based on threads. In *AAECC-8: Eighths Int. Symp. on Applied Algebra, Algebraic Algorithms, and Error-Correcting Codes*, volume 508 of *LNCS*, Tokyo, Japan, August 1990. Springer-Verlag.

[Küc91a] Wolfgang W. Küchlin. On the multi-threaded computation of integral polynomial greatest common divisors. In Stephen M. Watt, editor, *Proc. 1991 Internatl. Symp. on Symbolic and Algebraic Computation: ISSAC'91*, pages 333–342, Bonn, Germany, July 1991. ACM Press. (Also OSU-CISRC-1/91-TR2).

[Küc91b] Wolfgang W. Küchlin. On the multi-threaded computation of modular polynomial greatest common divisors. In *Proc. 1st Internatl. Conf. of the Austrian Center for Parallel Computation*, LNCS, Salzburg, Austria, October 1991. Springer-Verlag. (To appear).

[Loo76] R. G. K. Loos. The algorithm description language ALDES (Report). *ACM SIGSAM Bull.*, 10(1):15–39, 1976.

[Sam89] Ioannis Samiotakis. A thread library for a non-uniform memory access multiprocessor. Master's thesis, The Ohio State University, 1989.

[SLA89] B. D. Saunders, H. R. Lee, and S. K. Abdali. A parallel implementation of the cylindrical algebraic decomposition algorithm. In Gaston H. Gonnet, editor, *ISSAC'89*, pages 298–307, Portland, Oregon, July 1989. ACM-SIGSAM, ACM Press.

2

Algebraic Computing on a Local Net

Steffen Seitz[1]

Abstract:
An extension of the computer algebra system SAC-2 for the execution of algorithms by the workstations on a local net is described. We are more interested in the reduction of the execution time of existing algorithms than in the development of new parallel computing models. The semantics of the SAC-2 system language ALDES is extended for the specification of concurrent procedure calls. During the parallel execution of these calls the results are represented by futures, whose existence is hidden to the programmer by using data types. The current implementation, based on the SAC-2 I/O concept, is described. As an example for algorithms with many concurrent subtasks, the execution times for a distributed version of a modular algorithm are shown.

2.1 Introduction

The SAC-2 computer algebra system ([CL89]) is an open system. The complete source code of the system, that comprises over 500 algorithms, is available to the user. Though the system language ALDES is a high level language, it provides means for the detailed control of the execution of algorithms. The primary data type in SAC-2, that is supported by many algorithms and that is used as the means for representing algebraic objects, is the linked list. The notion of the algorithm is central in SAC-2. Extending SAC-2 means writing new algorithms, relying on existing ones, using them as elementary operations.

2.1.1 Overview

Our model is intended to help shorten the execution time of algorithms, which decompose a computation into several independent tasks. Then each task is assigned to a processor that is currently represented by one of several workstations on a local net.

The extension of ALDES with means for distributed and parallel computing

[1]Universität Tübingen, Wilhelm-Schickard-Institut für Informatik, Arbeitsgruppe Computer Algebra, Tübingen, Germany, Email: `seitz@informatik.uni-tuebingen.de`

should not change the property of SAC-2 to be an open system. The generation of multiple instruction streams, which always exist in parallel computing, should not be the consequence of the execution of code found by special phases during the compilation process, but it is the programmer who has to make the decisions by identifying promising points in algorithms, where parallel execution of code will lead to shorter execution times.

Without looking into program code the size of program segments, that can be executed efficiently in parallel, depends on the architecture of the underlying system. Because on multicomputer architectures message passing introduces overhead, it is not worthwhile to charge a processor with the execution of only a few instructions. Even if code is placed during the initialization phase, the time used for sending and receiving the arguments and the result exceeds the execution time. Thus decoupled processors require (dynamically) large program sizes. Similar to distributed systems running ARGUS [Lis87], *algorithms* are the smallest entities in the SAC-2 system, that are executable in parallel by the machines on a local net.

2.2 Model

A prerequisite for the parallel execution of a program, is the existence of concurrent code segments. Their independence can be used to execute them simultaneously. In our model, the programmer specifies procedure calls, whose execution is concurrent to the main program. *Futures* are used as preliminary results for result parameters after calling those procedures. The concept of futures has been introduced by [KHL76] for an extension of the ALGOL-68 language for the C.mmp multiprocessor and has been used by R. Halstead in MULTILISP ([Hal85]), by B. Liskov and L. Shrira in ARGUS ([LS88]) and by others. Because ALDES programs are intended to be compiled into machine executable code in order to get executed and the programmer should not be concerned with the existence of futures, they should be replaced by their value whenever the values are needed and not at positions, that have to be identified explicitly by the programmer.

2.2.1 Extending ALDES

In order to implement our model, two fundamental extensions have been added to SAC-2.

A more general interpretation of a SAC-2 device number allows the selection of programs as source or destination of data. Every character between peripheral devices and an ALDES program flows through the primitive operations `INPUT` and `OUTPUT` which have to be adapted during the installation process of SAC-2 to the underlying computer system. The unit of transfer between SAC-2 and operating system services is the *record*. Its size is determined by `ISIZE` for records to be read and by `RMARG` for records to be written. The device number, that denotes on single processor systems a peripheral device (e.g. a text file, the terminal or the printer), is used to denote any machine on a local net. The selection of a specific machine is currently done with the INTERNET address of the machine. The device number `HOSTS` in the variable `OUNIT` and the address in `OADDR` denotes the machine with address `OADDR` on the net. Because a machine is charged with at most one algorithm, a distinction between the processes running on that machine is not necessary. Incoming messages are held in a buffer, until they are read from the receiving ALDES program by calling `INPUT` with `IUNIT` set to `HOSTS` [2]. A similar extension to SAC-2, where every hosts gets a unique device number, has been described in [Sei89].

Example:

The polynomial P in r variables is written to the terminal with:

```
OUNIT:=6; ...; IPWRT(r,P); WRITE;
```

The following sequence writes it to (the buffer on) the host with address A (`SOADR` sets the address in A as the output address `OADDR` for `OUTPUT`:

```
OUNIT:=HOSTS; SOADR(A); ...; IPWRT(r,P); WRITE;
```

Reading the next object from the buffer, that is known to be a polynomial is done with:

```
IUNIT:=HOSTS; ...; IPRD(;r,P);
```

The character "`!`" is reserved as a prefix for concurrent procedure calls. It is intended to be interpreted by a preprocessor, that transforms concurrent procedure calls identified by the programmer into appropriate ALDES code for the remote execution of the call. The scheme is described in section 2.3.

The presentation of our model is guided by the following questions:

- How does the programmer specify concurrency?
- How is a concurrent procedure call executed?
- How is the existence of futures hidden from the programmer?
- How can the model be integrated into SAC-2?

2.2.2 Concurrency in ALDES

A procedure call with "`!`" in front of the procedure name, marks the call as concurrent to the further execution of the caller. In order to exploit this fact for executing program segments in parallel, a *task specification* is created as a result of the procedure call.

A task specification consists of the name of the called procedure, the values of the input parameters and the number of output parameters. Futures are used as values

---

[2] The buffer is used for *asynchronous* message passing. Various concepts of message passing are surveyed in [AS83].

for the result parameters and the execution continues, after the task specification has been sent to a scheduler.

When actual results are represented by futures, further execution of the program at the caller occurs. Values that may contain a future are called *incomplete*.

2.2.3 REPLACEMENT OF FUTURES

Because most operations require an ordinary value, futures have to be replaced by their value. This can be done in several ways:

1. The programmer includes code for the explicit replacement. Analogous to the creation of futures by marked procedure calls, there exists a procedure named GVF (=Get Value for Future) whose argument is a incomplete value. The result of GVF is the original value with all futures replaced by their proper value. In order to get every expression evaluated, on whose value the further execution depends, the programmer has to keep track of the content of every variable with incomplete values.

2. The code for the replacement of futures can be inserted by the compiler. The property of a variable, probably containing a future, can be described by an attributed grammar for the language. Whenever the compiler detects an expression E whose evaluation could lead to a future but the environment needs an ordinary value, E is replaced by GVF(E). This has not always to be the case for expressions used as arguments in procedure calls. The typical example is the procedure "cons". Cons works independent of the values of their arguments and may be called with incomplete values. The necessity of GVF can not be determined solely by a syntactical analysis of the program text. In the following example, after the first line the content of b may be a future. But if $h(a) \leq 0$ iff $f(a) > 0$ the insertion of GVF in order to be able to execute the negation is not necessary.

   ```
   if f(a) > 0 then b:=g(a) else b:=!g(a);
   ...
   if h(a) > 0 then c:=-b else c:=b;
   ```

   The automatic insertion of GVF during the translation process needs substantial modification to the ALDES translator.

3. In our model the replacement of futures is done implicitly by the usage of data structures. The data type IVB (=Incomplete Values, Bag) offers the following procedures:

```
                    B:=IVBNEW()
[ Incomplete Values, Bag, new
  An empty bag is created ]
```

```
                    B':=IVBADD(B,f)
[ Incomplete Values, Bag, add
  B is a bag of incomplete values, f an arbitrary
  value. f is added to B, the result is B' ]

                    IVBSEL(B;f,B')
[ Incomplete Values, Bag, select
  B is a bag of incomplete values. f is an ordinary
  value from B. B' is B without f ]

                    b:=IVBEMP(B)
[ Incomplete Values, Bag, is_empty predicate
  B is a bag of incomplete values, b is true if B is
  empty else b is false ]
```

The implementation exploits the fact, that members of a bag may be selected in arbitrary order. **IVBSEL** returns those members first, whose value is available.

The IVB type is used in a concurrent version of the SAC-2 resultant algorithm IPRES in chapter 2.4. In a first phase, all the futures for the necessary "image problems" are put into the bag. In the second phase the values for the image problems are selected from the bag for the final result.

2.3 Implementation

2.3.1 Machine Types

The machines on a local network are used for the implementation of the model by partitioning the machines into three classes. The user interacts at the *console* with his program in the same way, as he would do this in a single processor environment. Task specification packets, that are generated during the program execution, are sent to the *scheduler*. The scheduler sends them to one of several *algorithm servers* where the specified procedure will be executed with the supplied arguments. During the execution of procedures, new task specification may emerge from algorithm servers. Result packets are sent back to the initiator of a specification packet. During the initialization algorithm servers report the names of procedures, that can be executed by them to the scheduler.

2.3.2 Packets and Procedures for Distributed Computing

2.3.2.1 Packets

The various machines on the net communicate by sending and receiving *packets*. A packet is a distinguished message, sent from an ALDES procedures to an other machine in the same way as they would write external representations of objects to peripheral devices. Sending messages has been described as the first extension to SAC-2 in section 2.2.1.

The implementation of our model currently uses packets of the following types:

1. Specification packet
2. Result packet
3. Registration packet
4. Ready packet

In order to get a procedure call executed in parallel to the further execution of the calling procedure, a *specification packet* (SPECP) is sent to the scheduler. A SPECP for the execution of a procedure p with m input an n output parameters $(m, n \geq 0)$ consists of the following five parts:

1. The identifier "SPECP" (= the type of the packet)
2. An address (= the address of the sending host)
3. A name (= the name p of the procedure to be executed)
4. A list (= the list of m actual parameters)
5. A list of names (= the names of the n futures that represent the results)

In addition to SPECPs, the scheduler handles *registration packets* (REGPs). REGPs are sent from algorithm servers to the scheduler during their initialization. A REGP contains a value for the speed of the algorithm server and the names of the procedures it is able to execute. A REGP consists of the following parts:

1. The identifier "REGP" (= the type of the packet)
2. An integer value (= a performance value, high value means fast server)
3. A list of names (= the names of the procedures the server can execute)

The scheduler passes a SPECP to an algorithm server by looking at the name field in the SPECP and the list of procedure names reported by the servers. The SPECP is sent to an unused machine, that reported the highest performance value and that can execute the named procedure.

After the execution of a procedure, the server responds with the counterpart for a SPECP, a *result packet* (RESP). The RESP is sent to the host, that has emitted the SPECP. The content of result packets is the list of the futures associated with their values. A RESP consists of the following two parts:

1. The identifier "RESP" (= the type of the packet)
2. A list L of period 2 (= the list of m pairs "($future_i$ $result_i$)", $0 \leq i \leq m$")

In addition to sending a result, the server sends a *ready packet* (RDYP) to the scheduler, in order to tell its availability for further procedure execution. A RDYP consists of a single part:

1. The identifier "RDYP" (= the type of the packet)

2.3.2.2 Procedures

The following two procedures are used for generating and receiving specification and result packets. Sending SPECPs is done with the procedure STASP:

```
                    R := STASP(p, e, n)
[ Send Task Specification
  p is the name of a procedure with input arity m and
  output arity n, e is the list with length n,
  containing the actual input parameters. The task
  specification to apply p to e is sent to the
  scheduler. The result is the future or the list of
  futures for the result of applying p to e_1,...,e_m ]
```

STASP is intended to be used as the "meaning" of a concurrent procedure call. A call of the form

```
!p(e1,...,em;a1,...,an)
```

is finally executed as:

```
R:=STASP("p",(e1,...,em),n); ADVn(R;a1,...,an)
```

R is a new variable, not used in the current procedure. The list of n futures which form the result of STASP is spread by ADVn to the output parameters supplied with the call of p. When used with functions

```
!f(e1,...,em)
```

is compiled as

```
STASP("f",(e1,...,em),1).
```

Because the replacement scheme is easy it can be applied by a preprocessor that needs not much knowledge about the syntax of ALDES. Modifications to the original ALDES translator are not necessary.

The procedure GVF is used for the replacement of futures. A future is implemented as a *symbol* with the following properties:

1. ISREP : marks the symbol as a future
2. VALID : the value is known

3. VALUE : the value of the future
4. NEEDED : marks the value of the future as needed

The following procedure **GVF** makes the value of any future available, provided its computation terminated normally on the algorithm server. Because a server sends any result as a result packet to the sender of a task, **GVF** will either find the value for its argument at VALUE or it reads result packets until the desired value arrives. Result packets with unwanted values are read and the content is stored in the appropriate symbols for future calls of **GVF**.

```
const BUFFER=100.

                         v := GVF(f)
[ Get Value for Future
  f is a future, v is the value for f ]

(1) [init]  I:=IUNIT; IUNIT:=BUFFER.
(2) [get RESPs] while ~GET(f, VALID) do { repeat
    read until EOF # 1; if CREAD() # RESP then {
    print "Result expected."; DIBUFF; STOP };
    R := UREAD(); while R # () do { ADV(R;f',v',R);
    PUT(f', VALUE, v'); PUT(f', VALID, TRUE) } }.
(3) [requested value] v := GET(f, VALUE) ||
```

The programmer never uses **GVF** directly in his program. **GVF** is used for the implementation of data types which hide the existence of futures.

2.3.3 Passing Messages between ALDES Programs

The main purpose of distributed computation is the shortening of execution time, not the development of new, "distributed" algorithms. Therefore it is assumed, that at most one ALDES program, that participates in a distributed computation, is running on a machine at any time and the INTERNET address of a machine suffices as a unique destination address for messages. The extension of SAC-2 for sending messages is essentially the extension of the I/O primitives for reading and writing to and from *sockets* [Sec86], allowing the program to address a socket and thus any machine in the same way, as it would address the terminal or a file.

The ALDES program communicates with its own buffer and the buffer of other machines by *elementary packets*. Currently there are four types of elementary packets: REQUEST, REPLY, DATA, ACKNOWLEDGE. The structure of these packets has to be known only by the primitive operations that operate on the sockets and it is inaccessible on the ALDES program level. Elementary packets will not be described in further detail.

2.3.3.1 Asynchronous message passing

Because the language ALDES does not contain concepts for handling exceptions, the buffering of incoming messages is handled by a separate process, the "*buffer process*". The sending program delivers its messages to the buffer on the destination host and it continues. Thus the sender neither has to provide local buffer for undelivered messages nor it has to wait until the receiving ALDES program decides to read a message. Because SAC-2 organizes data streams as records, multiple records may be used to store the content of a single message.

An ALDES program reads messages by reading from the device `BUFFER`. After the input primitive `INPUT` has sent a request to the buffer, that it is awaiting the next record, one of the following two situations is possible:

1. The previous record sent was the last record of a message. The buffer responds with a `DATA` (elementary) packet, that either contains the head, i.e. the first record, of a new message and the value FALSE for EOF or EOF is set to TRUE, if the buffer does not contain any records.

2. The last record sent was an intermediate record of a message. The buffer immediately sends the next record or waits until the next record arrives before is responds with the `DATA` packet.

In order to recognize the last record of a message, messages have to be sent enclosed in parentheses. Opening and closing parentheses are the only character that are recognized by the buffer. A record is assumed to be the last record of a message, if it contains the last closing parentheses.

2.4 Example

2.4.1 IPRES

2.4.1.1 Algorithm

SAC-2 contains the modular algorithm `IPRES` for computing the resultant of two polynomials from $\mathbb{Z}[x_1, \ldots, x_n]$. `IPRES` successively computes the resultant of its arguments in fields $\mathbb{Z}_{p_i}$, $0 < i \leq m$, for which m is determinable in advance, and builds the result with the Chinese Remainder Algorithm. The n. resultant is computed after the $n-1$. has been integrated into the value for the final result.

Because it is know in advance, for which $\mathbb{Z}_{p_i}$ the resultant is needed, their computation can be started in a first phase, the results are collected in the second phase. The computations in the various $\mathbb{Z}_p$ can be done in parallel.

The "distributed" version `IPRES` in figure 2.1 uses the data type IVB for the representation of the concurrent computations.

The elements of the bag `T` are pairs of the form (p, C*). `p` is the name of the image and `C*` the future, that represents the resultant of the two polynomials `A` and `B` in $\mathbb{Z}_p$.

```
                         C:=IPRES2(r,A,B)
[Integral polynomial resultant, distributed version.
A and B are integral polynomials in r variables,
r ge 1, of positive degrees. C is the resultant of
A and B with respect to the main variable, an
integral polynomial in r-1 variables.]

    safe I,m,n,p,q.
(1) [Compute coefficient bound.] ...
(2) [Initialize.]  I:=PRIME;  Q:=1;  C:=0;  r':=r-1.
(3) [1. phase : Generate] t := 0; T := IVBNEW();
    while I # () /\ t = 0 do { ADV(I;p,I);
    A*:=MPHOM(r,p,A); if PDEG(A*)=m then {
    B*:=MPHOM(r,p,B); if PDEG(B*)=n then { C* :=
    !MPRES(r,p,A*,B*); T := IVBADD( T, (p,C*) );
    Q := IPROD(Q,p); if ICOMP(Q,f) >= 0 then t:=1 }
    }  }.
(4) [Algorithm fails.]  if I = () then {
    CLOUT("ALGORITHM IPRES FAILS."); EMPTOB; stop }.
(5) [2. phase : Collect] Q := 1;
    while ~IVBEMP(T) do { IVBSCT(T;t,T);
    FIRST2(t;p,C*); q:=MDINV(p,MDHOM(p,Q));
    C:=IPCRA(Q,p,q,r',C,C*); Q:=IPROD(Q,p) } ||
```

Figure 2.1. Distributed version of IPRES

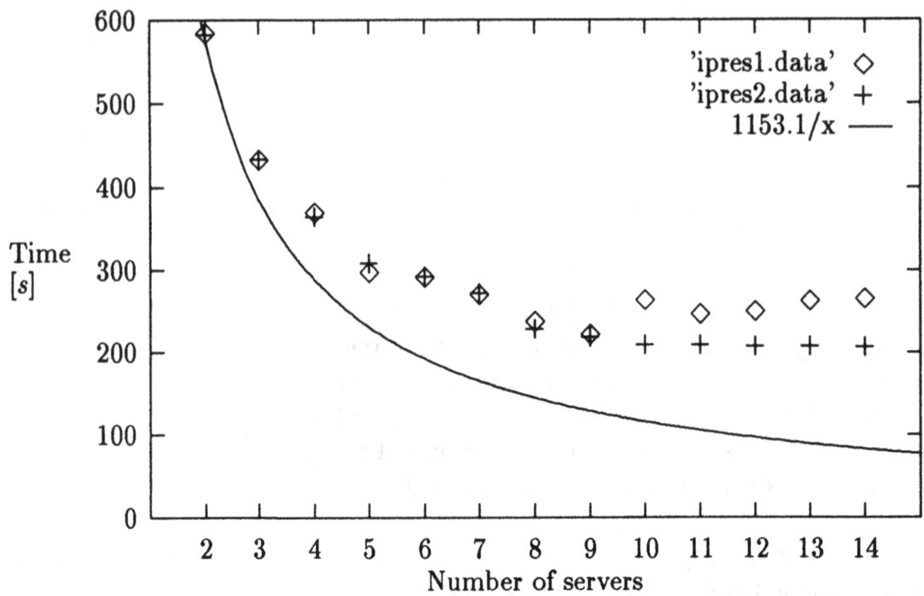

Figure 2.2. Runtimes for IPRES1 and IPRES2

2.4.1.2 Running times

The model has been implemented on a network of Sun workstations, running SunOS 4.0. Sun3/50 machines are used as algorithm servers. The scheduler runs on a Sun3/260, because it provides the speed for fast message handling.

Figure 2.2 shows the runtimes for the computation of the resultant of two polynomials from $\mathbb{Z}[x_1, x_2, x_3]$. The degrees for x_1, x_2 and x_3 is 3,4 and 5. During the computation, 14 task specification packets are sent from the console to the scheduler. The computation with the "old" single processor version `IPRES` requires 1153.1 s on a Sun3/50. The figure shows the times for the distributed versions of `IPRES` without the IVB data type (`IPRES1`, $\diamond$) and with IVB (`IPRES2`, $+$). In addition, the theoretical minimal runtime is shown as $y = \frac{1153.1}{x}$, with x the number of servers used.

The usage of two servers nearly halves the computation time in respect to the "old" sequential version `IPRES`. The further addition of servers reduces the computation time of `IPRES1` until 9 server are used. The reason for the *increase* of the computation time is as follows. The client (=`IPRES1`) asks for the value of specific futures and has to wait, until the appropriate result packet arrives. The results are expected to arrive in the same order as the task specifications have been generated. Because the times for the computations in `MPRES` are different, the results arrive in an arbitrary order. Although every incoming result could be used, they are only read and `IPRES1` *waits* for the arrival of the specific value.

The usage of the IVB routines in `IPRES2` avoids the waiting for specific results by using any available result. The runtime decreases until results from `MPRES` arrive in a higher rate at `IPRES2` than they are used for the construction of the final result in `IPRES2`.

2.5 Summary

The extension of the computer algebra System SAC-2 for the development of parallel algorithms suitable for the execution on a network of workstations has been described.[2] Minimal changes of existing algorithms are necessary in order to generate a distributed version of an existing sequential algorithm, whenever concurrent subtasks can be isolated. The administration of parallelism reduces, with the usage of appropriate data types, to a small amount. Because the programmer has very limited control over running subtasks, algorithms, which use speculative parallelism, i.e. which start computations before they are known to be necessary, are not supported by the model.

In addition, ALDES does not provide the means for the graceful handling of runtime errors. Currently the abortion of any process leads to the abortion of the whole computation. But this is a property of the implementation and not a property of the model.

Acknowledgements

I want to thank Prof. R. Loos for the supervision of my doctoral thesis, from which this paper is a part.

2.6 References

[AS83] G.R. Andrews and F.B. Schneider. Concepts and Notations for Concurrent Programming. *Computing Surveys*, 15(1):3–42, 3 1983.

[CL89] G. Collins and R. Loos. *SAC-2*, 1989.

[Hal85] R.H. Halstead. Multilisp: A Language for Concurrent Symbolic Computation. *ACM Transactions on Programming Languages and Systems*, 7(4):501–538, October 1985.

[KHL76] P. Knueven, P.G. Hibbard, and B.W. Leverett. A Language System for a Multiprocessor Environment. In *Proc. of the 4. Int. Conf. on the Design and Implementation of Algorithmic Languages*, pages 264–274, June 1976.

[2] The content of this paper uses several parts from my doctoral thesis, that appeared as technical report WSI–90–9 at the University of Tübingen.

[Lis87] B. Liskov. ARGUS Reference Manual. Technical report, Laboratory for Computer Science, Cambridge, Mass., 1987.

[LS88] B. Liskov and L. Shrira. Promises: Linguistic Support for Efficient Asynchronous Procedure Calls in Distributed Systems. In *Proceedings of the SIGPLAN 1988, June 22–24*, pages 260–267, 1988.

[Sec86] S. Sechrest. An Introductory 4.3BSD Interprocess Communication Tutorial. Technical report, DEECS, University of California, Berkeley, 1986.

[Sei89] S. Seitz. Parallel Algorithm Development. In J. Della Dora and J. Fitch, editors, *Computer Algebra and Parallelism*, chapter 15, pages 223–232. Academic Press, 1989.

3

An Environment for Parallel Algebraic Computation

Jean-Louis Roch[1]

Abstract:
PAC is a parallel environment, based on a MIMD distributed computing model, which is intended to aid in the development of computer algebra algorithms. It uses parallelism as a tool for processing large problems. This paper discusses the general relationship between computer algebra and parallelism. The general features of the PAC project are described and some of the results obtained with PAC are presented. One of the crucial elements of symbolic computation on parallel architectures is efficient implementation of fast arbitrary precision arithmetic. This paper presents a nodal (arbitrary precision) integer arithmetic package and discusses the fast division algorithm which we have implemented. The representation used is designed to take advantage of a vectorized floating point unit. Our experiences with this approach are also discussed.
PAC has been implemented on the FPS T series hypercube (32 processors)and an implementation on the TELMAT Meganode (128 processors) is in progress.

3.1 Introduction - Computer Algebra and Parallelism

Many algorithms in computer algebra divide large problems into a set of smaller problems which can be individually solved. This decomposition takes two forms: AND-parallel, where the results of all of the subproblems are combined and OR-parallel, where the first subproblem to obtain a solution is used in preference to all others. This section discusses each of these approaches.

3.1.1 AND-Parallelism in Computer Algebra

Consider a problem P chosen from a problem space E. One method that is often used to solve P is to project P onto sub-spaces of E, $E_1, \ldots, E_n$ so as to obtain sub-problems $P_1, \ldots, P_n$.

The solutions $s_1, \ldots, s_n$ of $P_1, \ldots, P_n$ may then be computed independently. Using a lifting process to combine the solutions $s_1, \ldots, s_n$, the solution s of the original

[1]Laboratoire LMC–IMAG, INPG–CNRS, 46, avenue Félix Viallet, 38031 Grenoble Cedex FRANCE, Email: `jlroch@mistral.imag.fr`

problem P may be computed.
Example 1: Assume P is a problem over the rationals numbers $\mathbb{Q}$, where the numerators and denominators of the solution coefficients are bounded by some known constant B, which usually depends on the problem P. Choosing several primes p_i such that $p_1 \cdots p_r > 2B$, solutions may be computed over $\mathbb{Z}/p_i\mathbb{Z}$, and then recombined to produce a solution over $\mathbb{Q}$. For instance, if P is the matrix inversion problem, this method, when applied to Gaussian elimination leads to a good solution [24].
Example 2: Manipulation of algebraic extensions is a basic problem in computer algebra. D. Duval [7] gives a nice approach to this problem, using a special strategy. Let μ represent a generic root of the polynomial P, which need not be irreducible. During computations with μ, it might become necessary to distinguish different roots of the polynomial P, in which case the problem splits into subproblems for each different cases. The results of these subproblems are then combined using AND-parallelism.

For example, let μ be a root of

$$P = x^3 - 2x^2 + x - 2 = (x^2 + 1)(x - 2),$$

(i.e., $\mu \in \{-i, i, 2\}$). When the code `if` $(\mu == 2)$ `then` $\{I\}$ `else` $\{J\}$ is executed two cases arise: if μ is a root of $P(x)/(x^2 + 1)$ then the code $\{I\}$ is to be executed. Otherwise, the code $\{J\}$ is to be executed. So the ring $E = \mathbb{Q}[x]/(x^3 - 2x^2 + x - 2)$ is split into two subrings:

$$E_1 = \mathbb{Q}[x]/(x - 2),$$
$$E_2 = \mathbb{Q}[x]/(x^2 + 1).$$

Thus the solution to the problem in E is viewed as the **union** of the solutions over E_1 and E_2.

3.1.2 OR-Parallelism in Computer Algebra

Probabilistic methods are a very valuable tool in Computer Algebra. They allow one to efficiently compute solutions, but with some small uncertainty in the validity of the results. These techniques are especially effective when there is a way to validate the probabilistic results. Although the expected time of a probabilistic algorithm is usually quite good, the actual time taken can be arbitrarily long with a suitably deviously chosen problem.

A good, predictable way of solving a problem for which there exists both a deterministic algorithm and an efficient probabilistic one is to use both algorithms in parallel and return the first (verified) solution that is returned.

This approach to computing obviously corresponds to the OR-parallelism approach. Examples of where this approach can be successful include the computation of polynomial GCD (where both efficient probabilistic and mush slower deterministic algorithms exist) and the computation of the standard basis using the Buchberger algorithm (using different strategies in parallel [22]).

3.2 The Parallel Model: General Overview of PAC

3.2.1 Adequation between Computer Algebra and Parallelism

The main reasons of intrinsically large complexity of computer algebra algorithms are:

- many elementary operations are performed: obviously, this is not typical of non-numerical problems. However, even small size problems, but badly conditioned ones, may let appear a large swelling in the size of intermediate coefficients, which increases the number of elementary machine operations to compute.

  Thus, MIMD machines, that allow us to perform different tasks in parallel seem well-suited to computer algebra algorithms.

- the objects involved are very large: to store such objects, a large memory space is needed. For example, Hermite normal form computation [23] is typical of this problem: even if coefficients of the solution are bounded, the bound on intermediate coefficients is enormous: thus, space needed for computation is very large compared with space needed for storing the result.

The genericity of distributed machines—and their small price too— seems attractive, as it allows the increase in local memory space needed for nodal computations. On such machines, the ability of associating to a given problem a dedicated topology is a very powerful tool. However, rooting data on a given topology may be a very difficult problem, if efficiency is required.

Another source of problems is the memory space available on each node: as we are considering power-extensible machines, this space is small (1 Mbyte on the physical models considered): so, the memory management on each node has to be carefully studied.

3.2.2 Target Architectures

A first version of PAC has been implemented on the hypercube FPS-T40 (32 processors T414). On this machine, several algorithms, based on a static distribution of tasks, have been studied. In this paper, we present some of the results.

A second version is being implemented on the TELMAT Meganode (128 processors T800). This new version allows dynamic distribution of tasks, with different levels of parallelism for a problem. This approach is based on the physical reconfiguration ability of this machine.

In both machines, the memory space available on each node is 1 Mbyte.

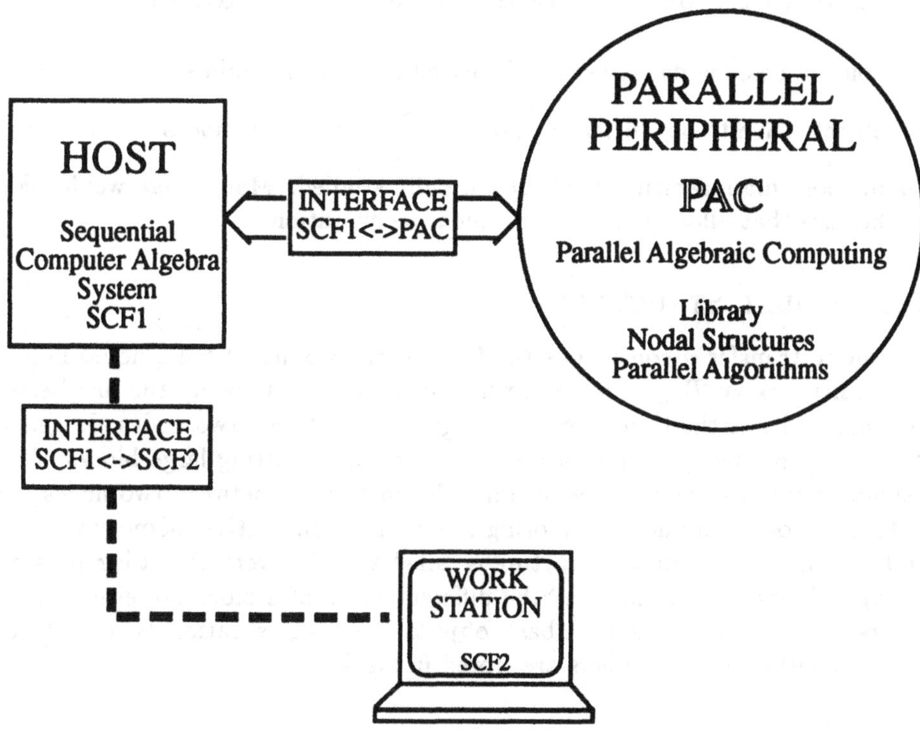

Figure 3.1. General Organization of PAC

3.2.3 General Presentation of PAC

PAC is an environment for computer algebra development, on a parallel distributed memory machine.

PAC includes:

- nodal primitives: on each node, several primitives are provided, that allow us to manage memory and to handle algebraic (symbolic) expressions.
- parallel primitives: suited to some expensive problems, they allow us to reduce these problems cost by using parallelism.

There are two ways of using PAC:

- as a basic library for development of parallel applications in computer algebra.
- as a specialized parallel unit, dedicated to the resolution of algebraic problems. PAC is interfaced with classical computer algebra systems (an interface with Reduce has been developed as a prototype). Thus, it may be easily used to solve, on a parallel machine, some specific problems whose sequential time cost—depending on the entries— is too expensive (polynomial arithmetic, linear system solving, normal form computations...).

3.3 Nodal Structures and Memory Allocation

The choice of basic nodal structures is driven by two constraints:

- objects have to be efficiently transmitted between two nodes
- memory management has to be carefully studied. Many users would like to be sure that allocation is a constant time operation.

3.3.1 Nodal Structures

The time to transfer n contiguous words between two neighboring nodes in a distributed network is: $T_{\text{com}} = \alpha + n\tau$, where α is the time to setup the link between two nodes and τ is the time to send a single word between two nodes. In general, $\alpha \gg \tau$. To minimize communication time when transmitting large blocks of data we should minimize the number of times the connection between two nodes is set up. This is most easily done by storing the data in consecutive memory words.

In PAC, this basic dynamic structure is called a *block*. Every object is represented as a typed block, of varying size: the different fields of a block are either data or pointers to other objects. All the basic objects like integers, rationals and arbitrary precision floating point numbers are stored in blocks.

3.3.2 Memory Management

Memory management is a crucial problem. If memory allocation is too expensive performance will be adversely affected. However, the best way of allocating memory is depends on the problem being solved, and more precisely, on the size of the blocks allocated during execution. In order to support a wide set of algorithms, PAC includes a number of different types of memory allocators from which the user may chose.

The simplest allocator is based on a free-list algorithm that guarantees (at least for applications requiring small space compared with the available physical memory) quick allocation (often constant) for most of the blocks. Logical or physical copies of a block are allowed.

However, for specific applications, this general-purpose approach is not sufficiently efficient. If basic objects of the problem have bounded size (for example, when modular arithmetic is involved) memory allocation may be performed more efficiently. For instance, J.L. Philippe [15] used PAC to implement the Pomerance quadratic sieve. The aim was to factor numbers less than 10^{200}. By developing a dedicated allocator—storing physically all numbers in blocks of constant size—performance was significantly improved.

3.4 Nodal Arithmetic

The implementation of rational arithmetic on the particular processors usded as nodes has been studied carefully [18].

The cost of addition on a sequential machine is linear in the size of the numbers being added. However parallelism does not improve integer addition because communications costs are also linear. Multiplication of integers, whose cost is assumed to be super-linear, is both theoretically (NC^1) and experimentally (even for small integers) easy to parallelize.

Division is more complex—it has not been proven to be in NC^1. The best known way to perform division, both in sequential or parallel, is based on a Newton iteration, in order to perform fast multiplications on same size numbers [4]. In the next subsection, we present an efficient way of computing this iteration so as to deal with large numbers of any size.

Integer GCD is very difficult to parallelize (not proven to be in NC). The best algorithm is only sublinear [10]. We have studied a generalization of the Lehmer algorithm, in order to take benefit of fast multiplications [18]; this algorithm is intrinsically sequential, but always faster than the Lehmer algorithm [13].

3.4.1 Division of any size integers

The block $[a_n, a_{n-1}, \ldots, a_0]_\beta$ with $0 \leq a_i < \beta$ and $a_n \neq 0$ if $n > 1$ represents the integer

$$a = \sum_{i=0}^{n} a_i \beta^i.$$

The a_i are called β-digits.

S.A. Cook [4] proved that euclidean division of two integers $u = [u_{2n+1}, \ldots, u_0]_2$ and $v = [v_n, \ldots, v_0]_2$ (in radix $\beta = 2$) is reducible to multiplication.

Let $M_\beta(n,p)$ denote the complexity of multiplication of two integers with n and p β-digits, respectively. Here, we are interested in a bound for the complexity $D_\beta(n,p)$ of division of two arbitrary integers, with n and p β-digits, $n \gg p$, respectively.

We propose a generalization of the computation with reciprocal introduced by Cook. A special trick allows us to efficiently compute the Newton iteration, and leads to a fast practical implementation.

3.4.1.1 Introduction

The α-reciprocal in basis β of an integer v is the integer $\widetilde{\rho_\alpha}$ defined by:

$$\widetilde{\rho_\alpha} = \left\lfloor \frac{\beta^\alpha}{v} \right\rfloor \tag{3.1}$$

Let $u = [u_p, \ldots, u_0]\beta$ and $v = [v_n, \ldots, v_0]_\beta$ be two integers, with $v_n \neq 0$ and $u_p \neq 0$. Clearly we have

$$\left\lfloor \frac{u}{v} \right\rfloor = \lfloor u \times \widetilde{\rho_{p+1}} \times \beta^{-p-1} \rfloor + c, \quad \text{with } c \in \{0, 1\}$$

The choice $\alpha = p + 1$ is optimal in the sense that the pth β-digit of v^{-1} directly concerns the value of q. Besides, only the $(p - n)$ first β-digits of ρ_α are to be computed, as the n first β-decimals of v^{-1} are zero.

The basic iteration used to compute $\widetilde{\rho_{p+1}}$ is:

$$\rho_{k+1} = 2\rho_k - \rho_k^2 \frac{v}{\beta^{k+1}} \tag{3.2}$$

However, on the one hand the values ρ_k computed in (3.2) are rationals, although $\widetilde{\rho_\alpha}$ is an integer, and on the other hand, the number of β-digits of $\widetilde{\rho_{p+1}}$ (both right and unfortunately wrong digits) is doubled at each step. To avoid this problem, two kinds of algorithms are used:

- computing (3.2) in $\mathbb{Q}$ until a certain number of good β-digits is obtained, and then truncating the obtained value to give $\widetilde{\rho_k}$ [4] [11]
- computing (3.2) in $\mathbb{Z}$ and correcting ρ_k at each step so as to obtain $\widetilde{\rho_k}$ [1]. The cost of this algorithm is:

$$C(n, p) = 5M_\beta(n - p, n - p) + O(n + p) \tag{3.3}$$

This cost may be improved when $n > 2p$: the algorithm presented here takes benefit of the "leading" zero digits in the divisor to compute ρ_{k+1} (3).

3.4.1.2 An algorithm for fast division of arbitrary size integers

The aim of the algorithm is an efficient way of computing (3.2) to obtain (3.1) when $n \gg p$.
Initialization: let r_0 be the reciprocal of v:

$$\widetilde{\rho_0} = \left\lfloor \frac{B_0}{v} \right\rfloor \quad \text{with } B_0 = \beta^{p+1} \tag{3.4}$$

Let $B_{n+1} = B_n^2$ and $\widetilde{\rho_n} = \lfloor B_n / v \rfloor$ is assumed to be already computed. We compute $\widetilde{\rho_{n+1}}$ with (3.2):

$$\widetilde{\rho_{n+1}} = 2\widetilde{\rho_n} B_n - \widetilde{\rho_n}^2 v \tag{3.5}$$

Let c be the correction so as to obtain

$$\widetilde{\rho_{n+1}} : \widetilde{\rho_{n+1}} = \rho_{n+1} + c \tag{3.6}$$

c can be deduced from the following equation:

$$c = \left\lfloor \frac{B_n^2 - \rho_{n+1} v}{v} \right\rfloor \tag{3.7}$$

Then c and ρ_{n+1} may be eliminated from (3.6) by applying (3.5) and (3.7). Thus, by decreasing the dividend involved in the division by v (3.7), the final equation (3.8) is obtained:

$$\widetilde{\rho_{n+1}} = \widetilde{\rho_n} B_n + \Delta_n \rho_n + \left\lfloor \frac{\Delta_n^2}{v} \right\rfloor \tag{3.8}$$

where Δ_n is defined by:

$$\begin{aligned} \Delta_0 &= B_0 \pmod v \\ \Delta_{n+1} &= \Delta_n^2 \pmod v \end{aligned} \tag{3.9}$$

Δ_{n+1} and $\lfloor \Delta_n^2 / v \rfloor$ are easily computed using a multiplication with $\widetilde{\rho_0}$ reciprocal of v, computed in (3.4). At each step, the important following equalities are verified by Δ_n and $\widetilde{\rho_n}$:

$$\forall n. \widetilde{\rho_n} = \left\lfloor \frac{B_n}{v} \right\rfloor \quad \text{and} \quad \Delta_n = B_n \pmod v. \tag{3.10}$$

This proves $\widetilde{\rho_n}$ to be a generalized reciprocal of v.

The complexity of the corresponding algorithm is:

$$C'(n,p) = \left(5 + \frac{n+1}{2p-1}\right) M_\beta(p,p) + O(n+p) \tag{3.11}$$

which improves the previous result (3.3).

The complexity of the algorithm for euclidean division may be deduced from (3.11):

$$D_\beta(m,p) = \left(5 + \frac{n+1}{2p-1}\right) M_\beta(p,p) + M_\beta(n, n-p) + M_\beta(p, n-p) + O(n+p).$$

3.5 Vectorization of Integer Arithmetic

3.5.1 Introduction—Model

As integer arithmetic involves only basic processor operations, parallelizing this arithmetic is experimentally interesting only when operands are very large. That is why we turn to another model of fast computation: the use of vector process units.

The chosen model includes:

- one control unit (which decodes one instruction at each cycle)
- one arithmetic unit, including one adder and one multiplier for floating point numbers (IEEE64). Both units are pipelined, and contain respectively e^+ and $e^\times$ stages.

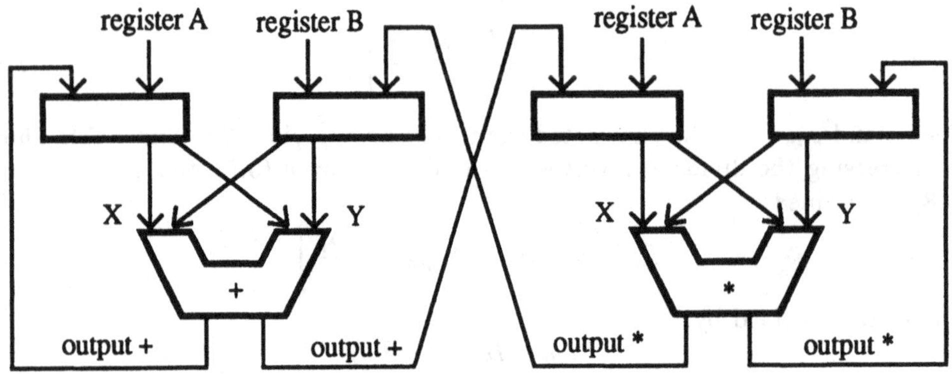

Figure 3.2. Vector Process Unit: Model

- several vector registers, all of size T_r, with direct access to both pipelines.
- two scalar registers—one for each arithmetic unit.
- the memory is split into two banks, with direct access to registers.

The vector unit is assumed to perform an operation in three steps:

1. Loading operands, scalar registers, operations code;
2. Computation (the size of vectors is given);
3. the result is stored in one register;

This model corresponds to most of vector processing units. Experiments have been performed on a Weitek VPU, for which $e^{+} = 5$, $e^{\times} = 6$, $T_r = 128$.

3.5.1.1 Redundant representation

Standard algorithms for integer addition and multiplication are well suited to vectorization: however, it is fundamental to avoid carry propagation. So, we introduce a dedicated redundant representation: let Δ and M be two positive integers.

The sequence $\{ u_n, \ldots, u_0 \}_{\Delta,M}$ (with $-M \leq u_i \leq M$) denotes the integer U:

$$U = \sum_{i=0}^{n} u_i \Delta^i \tag{3.12}$$

Δ is chosen such that a product of two integers of T_r digits prevents carry propagation: this is realized if and only if $(T_r - 1) \times (\Delta - 1)^2 \leq M$ i.e.:

$$\Delta \leq 1 + \sqrt{\frac{M}{T_r - 1}} \tag{3.13}$$

Algorithm	Experimental cost
Non vectorized	$0.020n^2 + 0.37n + 0.55$
vectorized	$0.0055n^2 + 1.2n + 5.2$
Karatsuba splitting	$0.090n^{1.58}$

Figure 3.3. Experimental cost of integer multiplication algorithms

M is set to the greater positive integer exactly represented in the floating point number format. If m is the number of bits of the mantissa, we thus have:

$$M = 2^m - 1 \tag{3.14}$$

Application to the experimental model: $T_r = 128$, $m = 53$ (IEEE64). (3.13) and (3.14) lead to the value of M and Δ: $M = 2^{53} - 1$ and $\Delta = 2^{16}$ (Δ is a power of two, more suitable to the communication with memory-registers; this choice makes conversions with β-basis representation easier.

With this representation, integer addition corresponds to vector addition, and integer multiplication to n SAXPY operations on vectors (n is the number of digits of the smaller operand).

To ensure the representation (3.12) does not overflow, we introduce a control primitive, denoted $|\ |_{\Delta,M}$, easy to evaluate:

$$|\{\, u_n, \ldots, u_0 \,\}_{\Delta,M}|_{\Delta,M} = \max\{|u_i|\}_{i=0,\ldots,n}.$$

Let $U = \{\, u_n, \ldots, u_0 \,\}_{\Delta,M}$ and $V = \{\, v_p, \ldots, v_0 \,\}_{\Delta,M}$: $S = U + V$ and $P = UV$ are to be computed.

If $|U|_{\Delta,M} + |V|_{\Delta,M} < M$ then S may be computed with the vector adder, and : $|S|_{\Delta,M}$ is set to $|U|_{\Delta,M} + |V|_{\Delta,M}$

If $\inf(n,p)|U|_{\Delta,M}|V|_{\Delta,M} < M$ then P may be computed with the vector adder and multiplier: $|P|_{\Delta,M}$ is set to $\inf(n,p)|U|_{\Delta,M}|V|_{\Delta,M}$

If these conditions are not true, then U and V have to be normalized so as to obtain a representation where coefficients u_i and v_i are positive and bounded by Δ. This normalization consists only in one carry propagation through the number.

3.5.1.2 Cost of integer multiplication on a vector unit—Experimentation

The cost of the standard multiplication (several SAXPY) is then:

$$T_{\text{mul}}(n,p) = n \times (p+1) + (p+1) \times e^{\times} + p \times e^{+} \quad \text{cycles}$$

So, vectorization should be very interesting: however, on the one hand the linear term in the cost makes the use of the vector unit very expensive for small integers

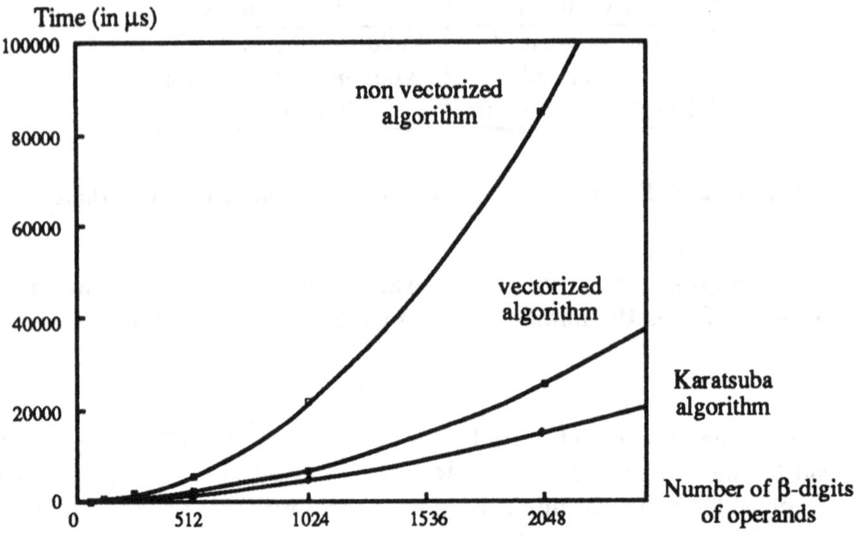

Figure 3.4. Performances of algorithms which multiply large integers

(< 40 β-digits), and on the other hand, the order of the standard algorithm makes it slower than fast multiplication algorithms for large integers (> 30 β-digits).

It is very important to notice that the magnitude of the constant term is due to the chosen unit: on higher performance units, loading registers is faster.

3.6 How to Take Benefit of Distributed Parallelism

3.6.1 Analysis of the Problem and Parallel Algorithms

The sequential algorithm consists of generating monomials of the product by decreasing order in degrees—using a heapsort structure—[8]. A direct parallelization consists in splitting one of the two operands; then two balanced products may be obtained [19]. Let

$$P = \sum_{i=0}^{p} a_i X^{\alpha_i} \quad \text{and} \quad Q = \sum_{i=0}^{q} b_i X^{\beta_i}$$

be two polynomials. The computation scheme is described in Figure 3.5.

This parallelization is well-suited to hypercube topology. At step i, the processors in dimension lower than i split their operands, and distribute half of their tasks to their neighbor in dimension $i+1$ (Figure 3.6).

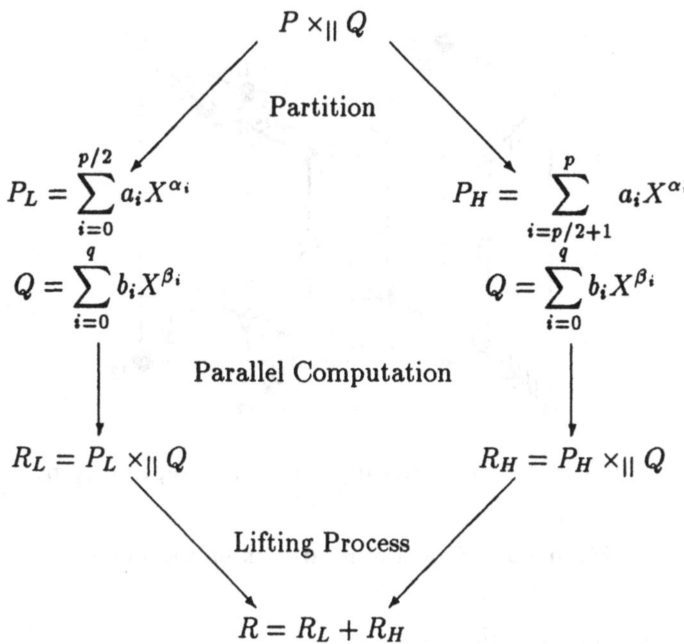

Figure 3.5. Parallel algorithm computation scheme

3.6.2 Choice of the polynomial to split

This choice is related to two constraints. On the one hand, by splitting the smaller polynomial, less arithmetic operations have to be computed. On the other hand, so as to minimize length of communicated data, it is better to split the larger polynomial.

3.6.3 Complexity study

Coefficients of polynomials are assumed to be bounded by B, and polynomials are sparse with respect to each other (the number of monomials of their product is the sum of the numbers of monomials of each polynomial).

With the previous conventions, times needed for the sequential and parallel algorithms are [18]:

$$\begin{aligned} T_{\text{seq}}(P,Q) &= pqT_{\text{mul}}(B(P),B(Q)) + (p,q-1) \times T_{\text{add}}(B(P)+B(Q)) \\ T_{\text{par}}(P,Q) &= pq\left(\frac{T_{\text{mul}}(B,B)}{2^N} + T_{\text{add}}(B,B) + T_{\text{com}}(B)\right). \end{aligned} \tag{3.15}$$

If p and q are small compared to 2^N, the parallelization is unefficient. Let us consider the computation of a large product, assuming that p and q are very large compared to 2^N.

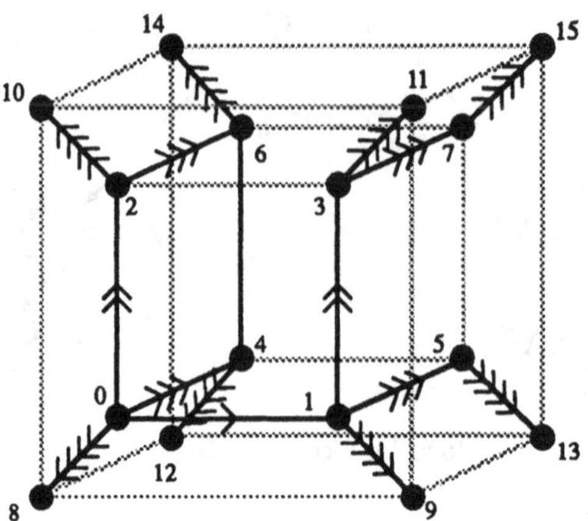

(The number of arrows indicates the reception step)

Figure 3.6. Graph of repartition on the 4-cube

Then, the corresponding efficiency is:

$$e = \frac{T_{\text{seq}}}{2^N T_{\text{par}}} = \frac{T_{\text{mul}}(B,B) + T_{\text{add}}(B,B)}{T_{\text{mul}}(B,B) + 2^N(T_{\text{add}}(B,B) + T_{\text{com}}(B))}$$

So, if B is large enough, communication costs become negligible, and non-linear coefficient multiplication costs are the largest: then, efficiency is close to one.

3.6.4 The best number of processors

The compromise between number of processors and size of the problem for which parallelism is of interest, may be here precisely studied.

The parallelization is of no benefit as soon as the number of processors 2^N is larger than 2^{N_0}, where N_0 is defined by:

$$T_{\text{Par}}^{N_0-1}(P,Q) > T_{\text{Par}}^{N_0}(P,Q)$$

$$T_{\text{Par}}^{N_0}(P,Q) \leq T_{\text{Par}}^{N_0+1}(P,Q)$$

Assuming N_0 is large enough ($2^{N_0} \gg 1$), from (3.15) we may obtain the following approximation of N_0:

$$N_0 \geq \log_2\left(1 + \frac{2p(B_p + B + q)g(B + p + B_q)}{\tau B_q + \alpha}\right).$$

3.6.5 Experimentation and Conclusion

Other algorithms for this problem have been studied, especially Karatsuba splitting [18]: we compare here the experimentally obtained results.

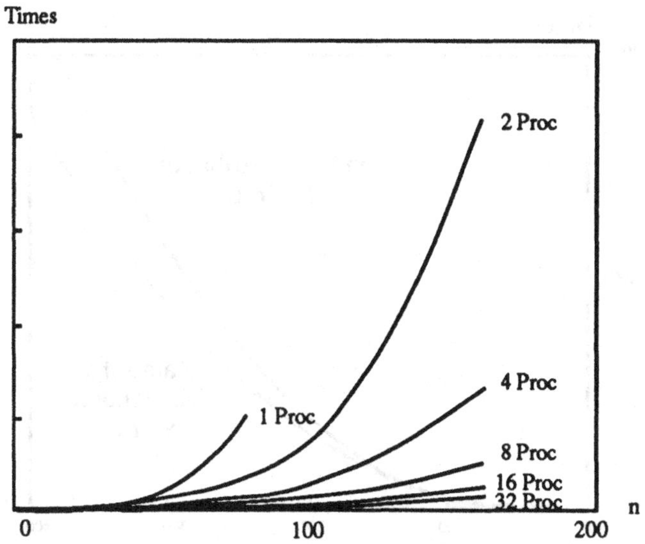

Figure 3.7. Parallelization of Johnson Algorithm, $P = Q = (x + y)^{5n}$

For both algorithms, implementation shows that the best choice is to split the larger polynomial, so as to decrease communication times: this is justified by the ratio (communication time)/(computation time) on the FPS-T40. This conclusion is still valid on the Meganode.

It is very difficult to study practically efficiency: by increasing the number of processors, larger problems may be solved (mainly because of the corresponding memory extension). However, speed-up is good, even if, in the benchmark used here, parallel tasks are unbalanced (middle coefficients are much larger than extremal ones). In order to take in account this increase in the size of the treated problems, an extension of Gustafson' speed-up has be studied [23].

3.7 Some Applications in PAC

We present here some results obtained within the PAC project on the FPS T20 and T40 [19].

3.7.1 Exact Solution of linear systems with rational coefficients

This problem has been studied by Gilles Villard [24]. Different algorithms have been implemented: standard Gauss inversion, Gaussian elimination using p-adic developments or modular arithmetic.

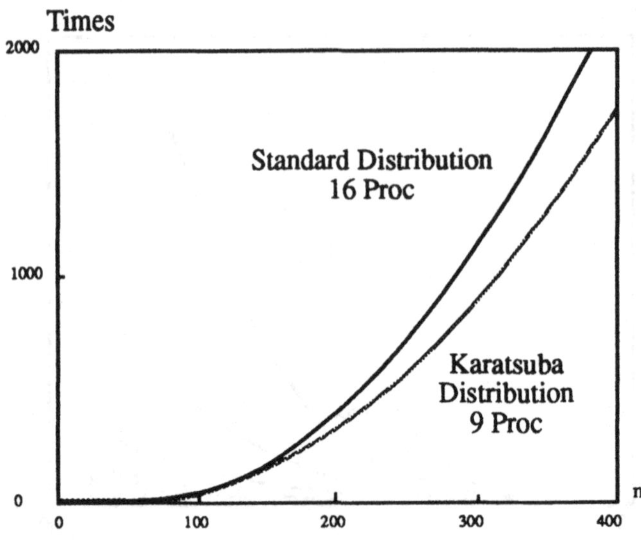

Figure 3.8. Standard versus Karatsuba Distributions, $P = Q = (x + y)^{5n}$

The implementation allows us to solve 700 × 700 dense system—with integral coefficient in entry bounded by 100—in 3 hours 35 minutes on 16 processors.

3.7.2 Hermite Normal Form Computation on integers and polynomials

This problem has been studied by Françoise Siebert-Roch [23]. This problem has the particularity that it needs a large memory space to be computed.

Two cases are distinguished. In the integral case, the llopoulos algorithm gives the best results: two ways of distributing data have been studied (by rows and columns [23]). The implementation allows us to compute Hermite form of a 160 × 160 matrix in 3 hours on 32 processors: for this computation, 5 Gbytes of memory were allocated.

In the polynomial case, a parallelization of the Kannan algorithm has been proposed. However, the obtained speed-up is not as good as in the integral case, because of the inherent unbalancing of the tasks: only a dynamic approach would bring efficiency to this peculiar problem, as it is impossible to balance statically the elementary tasks because of the swelling of intermediate coefficients.

The parallel computation of the Hermite normal form of a matrix (using PAC) is interfaced with the Reduce computer algebra system, and so may be called directly from this system.

3.8 Conclusion

The results obtained illustrate the usefulness of parallelism in computer algebra. The validation of theoretical algorithms with an implementation on a dedicated model is very important.

The implementation in progress on the Telmat-Meganode makes possible the use of quite a large number of processors. The use of different levels of parallelization becomes interesting as it allows us to dynamically take advantage of parallelism.

Parallelism and computer algebra seem to be strongly related. On the one hand, the modelization of parallel And and Or mechanisms with an algebraic point of view should bring powerful tools to the description of computer algebra algorithms. On the other hand, different applications of computer algebra give a new point of view to parallelism: one of the most important is the straight-line program parallel evaluation [14].

3.9 Acknowledgements

This work is supported by the *PRC Mathématiques et Informatique* and by the *GRECO Calcul Formel* of the French *Centre National de la Recherche Scientifique (CNRS).*

3.10 References

[1] A.V.Aho, J.E.Hopcroft, J.D.Ullman *The Design and Analysis of Computer Algorithms*, Addison Wesley (1974)

[2] A. Borodin, "On relating Time and Space to Size and Depth", *SIAM Journal of Computing*, 5, pp. 733–744, (1977).

[3] R. Cole, U. Vishkin, "Optimal Parallel Algorithms for Expression Tree Evaluation and List Ranking", Springer Verlag, Lectures Notes in Computer Science, 319, pp 91–100 (1988).

[4] S.A.Cook, "On the Minimum Complexity of Functions", *Trans. Amer. Math. Soc.*, 142, pp. 291–314 (1969).

[5] S.A. Cook, "A Taxonomy of Problems that have Fast Parallel Algorithms", *Information and Control*, vol. 64, pp. 2–22, (1985).

[6] L. Csanky, "Fast Parallel Matrix Inversion Algorithms", *SIAM Journal of Computing*, 5/4, pp. 618–623, (1976).

[7] C. Dicrescenzo and D. Duval, "Algebraic Extensions and Algebraic Closure in Scratchpad II," *Proceedings of ISSAC'88, (Ed. P. Gianni), Lecture Notes in Computer Science 358*, Springer-Verlag, New York, 1989.

[8] S. Johnson "Sparse Polynomial Arithmetic", Bell Labs research report (1974).

[9] A. Karatsuba, Y. Ofman "Multiplication of multidigit numbers on automata", *Dok. Akad. Nauk. SSSR*, vol.145 (p. 293–294) (1962)

[10] R.Kannan, G.Miller, L.Rudolph, "Sublinear parallel algorithm for computing the gcd of two integers", *SIAM J. Computing*, vol 16/1, February (1987).

[11] D.E.Knuth, *Semi-numerical algorithms*, Addison-Wesley (1981).

[12] S.R. Kosaraju, V. Ramachandran, "Optimal Parallel Evaluation of Tree-Structured Computations by Raking",, Springer Verlag, Lectures Notes in Computer Science, 319, pp. 101–110 (1988).

[13] D.H.Lehmer,"Euclid's algorithm for large numbers", *AMMM*, 45, pp. 227–233, April (1938).

[14] G.L. Miller, E. Kaltofen, V. Ramachandran, "Efficient Parallel Evaluation of Straight-Line Code and Arithmetic Circuits", *SIAM Journal of Computing*, 17 / 4 pp. 687–695 (1988).

[15] J.L. Philippe, *Parallélisation du crible quadratique—Application à la cryptographie*, Ph.D. Thesis, Inst. Nat. Polytechnique de Grenoble (1990).

[16] N. Pippenger, "On Simultaneous Resource Bounded Computation", *Proc. 20th Ann. IEEE Symp. on Fund. of Computer Science*, pp. 307–311, (1979).

[17] N. Revol, "Evaluateur distribué d'expressions arithmétiques", Rapport de fin d'études, Inst. Nat. Polytechnique de Grenoble (1990).

[18] J.L. Roch, *Calcul Formel et Parallélisme: Le système PAC et son arithmétique nodale*, Ph.D. Thesis, Inst. Nat. Polytechnique de Grenoble (1989).

[19] J.L. Roch, P. Sénéchaud, F. Siebert, G. Villard, "Computer Algebra on MIMD machine", *SIGSAM Bulletin*, 23/1, January (1989).

[20] W.L. Ruzzo, "On Uniform Circuit Complexity", *Journal of Computer and System Sciences*, 22, 3, pp. 365–383, June (1981).

[21] Y. Saad, M.H. Schultz, "Topological Properties of Hypercubes", *IEEE Transactions on Computers*, 37/7 (1988).

[22] P. Sénéchaud, *Calcul Formel et Parallélisme: Bases de Gröbner Booléennes—Méthodes de Calcul—Applications—Parallélisation*, Ph.D. Thesis, Inst. Nat. Polytechnique de Grenoble (1990).

[23] F. Siebert-Roch, *Calcul Formel et Parallélisme: Forme Normale d'Hermite—Méthodes de Calcul et Parallélisation*, Ph.D. Thesis Inst. Nat. Polytechnique de Grenoble (1990).

[24] G. Villard, *Calcul Formel et Parallélisme: Résolution de systèmes linéaires*, Ph.D. Thesis, Inst. Nat. Polytechnique de Grenoble (1988).

[25] H.J. Yeh, "L'intepréteur PAC, et l'évaluation parallèle dynamique", Rapport de fin d'études, Inst. Nat. Polytechnique de Grenoble (1990).

4

Finite Field Arithmetic Using the Connection Machine

Ernest Sibert[1]
Harold F. Mattson[1]
Paul Jackson[2]

Abstract: A Connection Machine (model CM-2) with 32K processors has been used to carry out calculations in finite fields with as many as 2^{21} elements and of various characteristics; a typical calculation is to determine the number of roots of a large family of polynomials. The programs use discrete logarithms, employing a table of "successor" logarithms to perform addition. The table is computed in advance, in parallel. The system can evaluate some 4×10^6 polynomial terms per second; performance is limited by the general communication time needed for table lookup. Orbits of the p-th power bijection (also calculated in parallel) are used to deal with common symmetries arising in the calculations. The techniques are illustrated by calculations to determine the number of rational points of a polynomial surface over several fields, quantities which are useful in analyzing certain cyclic codes.

4.1 Introduction

The recent introduction of massively parallel, SIMD computers such as the Connection Machine[3] makes it feasible, indeed easy, to carry out some calculations which only a few years ago would have been at best costly, at worst prohibitively so. Suppose, for example, we wish to determine the roots of a polynomial $f(x)$ over a finite field F. A very simple, but sensible, approach using a parallel computer with as many processors as the number of elements of F is:

- Calculate $f(x)$ for all $x \in F$ in parallel
- Identify (or count) the roots in parallel

For fields of no more than a few million elements we expect to perform such a calculation in a small fraction of a second. More importantly, we can obtain useful

[1]CIS - 44–116 CST, Syracuse University, Syracuse, NY 13244–4100
[2]Northeast Parallel Architectures Center, 3-201 Center for Science and Technology, Syracuse University, Syracuse, NY 13244-4100
[3]Trademark of Thinking Machines Corporation, Cambridge, Massachusetts

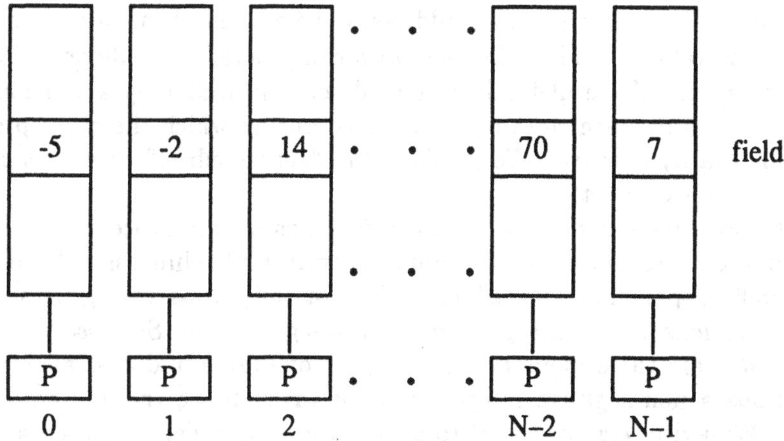

Figure 4.1. Array of processors and memories

information about the roots of a sizable family of polynomials in a reasonable time. Calculations of the latter kind arise, for example, in analyses of cyclic codes [4]. As a first step in exploring this possibility we have written programs to perform such calculations on a Connection Machine Model CM-2 [1, 8] with 32K (2^{15}) processors, using the *Lisp programming language[3] [7].

4.2 The Connection Machine

Connection Machines (model CM-2) are manufactured in sizes ranging from 4K to 64K processors (1K = 1024) with memories from 64K bits per processor to 1M bits per processor (1M = 1024K = 2^{20}). For example, the system used in this study has four quadrants of 8K processors each (64K bit memories), and may be operated as one machine of 32K processors (and memories) or as independent machines of 8K or 16K processors. A Connection Machine [CM] is not itself a stored-program computer; rather it is controlled by a conventional computer known as a *host* (a Vax 8800[4] in our system). A CM program is executed by the host, generating a stream of commands to be obeyed by the CM. Results of the parallel computation in the CM may be returned to the host or transferred directly to graphic displays or special mass-storage devices.

By a *field* of the CM memory (no relation to the algebraic structure of the same name) we mean a set of blocks of contiguous bits beginning at the same address in each memory and having a specified length. (See figure 4.1, CM memory addresses select individual bits.) A representative arithmetic operation has the form $a := b+c$, where a, b, c denote appropriate memory fields, and is performed simultaneously by

[4]Trademark of Digital Equipment Corporation, Maynard, Massachusetts

all processors. While the memory address and field length are the same for all processors, the values found in the memories will generally be different. Programs may, as the result of parallel calculation, determine that only a portion of the processors should perform certain instructions, during which the other processors remain temporarily inactive. We speak of the "active subset" of processors or of the "currently selected set".

The CM provides a variety of specialized means of interprocessor communication, including nearest-neighbor communication in multi-dimensional grids, scans (parallel prefix), spreading, and reduction, but for our purposes the general communication operations 'send' and 'get' are of most significance. Suppose that *source*, *dest*, and *addr* are three memory fields. Think of *source* and *dest* as data fields; *addr* contains a non-negative integer which is taken to be the address of a CM processor. We write, *e.g.*, *source*[i] to denote the *source* field in processor i. The send operation copies *source*[i] into *dest*[*addr*[i]] for each active processor i. The processors into whose memories values are written may be active or inactive. If two or more values might be sent to the same destination we can specify one of several combining operations to be used with send, *e.g.*, send-with-add, in which case the sum of the values sent will be left in the destination. The get operations copies *source*[*addr*[i]] to *dest*[i] for each active processor i. Here the processors from which values are read may be active or inactive, and many processors may read from the same location.

Although the details are not particularly relevant, we should mention that communication is implemented using a hypercube network whose nodes are clusters of 16 CM processors and memories. It will come as no surprise that general communication is much slower than other operations. Where basic arithmetic might require 20-30 microseconds (μsec.), a send could involve 2-3 milliseconds (msec.), and get can require 6-8 msec. Operation times are not constant, however: arithmetic times depend on field lengths and the nature of the operation, communication times vary enormously depending on address patterns and the number of active processors.

The CM also provides a "virtual processor" mechanism which permits emulation of a machine with many more processors than the physical array, though with correspondingly smaller memories. This is a considerable convenience, since it lets us write programs based on the assumption that the machine configuration fits the calculation to be performed, whatever the underlying hardware. The virtual processor mechanism is quite efficiently implemented as well (in hardware and microcode). Using 1M virtual processors on 32K physical processors (a "vp ratio" of 32) we find that typical operations take not quite 32 times as long as when using the 32K physical processors directly. Communication operations might not always perform so well at higher vp ratios, but in favorable cases they do.

4.3 Computational Approach

The customary approach [5] represents elements of GF(q) as vectors over its prime subfield, perhaps using a polynomial basis. Addition and subtraction are performed directly in the vector representation, multiplication and division are performed by means of discrete logarithms, using precomputed tables of logarithms and antilogarithms. Here we use discrete logarithms, but avoid the vector representation for reasons that will shortly become apparent.

We write **0** for zero in GF(q), **1** for the unit in GF(q), $l(x)$ for the discrete logarithm of $x! = \mathbf{0}$, corresponding to some primitive element b of GF(q). Thus $l(b^i) = i$ for $0 \leq i < q-1$. We represent non-zero elements of GF(q) by their discrete logarithms, $0, \ldots, q-2$; **0** is represented by the artificial "logarithm" $q-1$ (or some other convenient value), and we adopt the convention that $l(\mathbf{0}) = q-1$. We suppose that our CM has at least q virtual processors and store a "successor table" s as a field in the CM memory (one entry per processor) calculated in advance so that

$$s[i] = l(b^i + \mathbf{1}) \qquad \text{if } i < q-1$$

$$s[q-1] = 0 = l(\mathbf{1}).$$

Note that the successor entry is itself a logarithm, and satisfies $b^{s[i]} = b^i + \mathbf{1}$ for $i < q-1$. How the successor table is generated will be discussed a little later. With all this, we can implement the basic field operations for the case $x! = \mathbf{0}$ and $y! = \mathbf{0}$ from the equations:

$$l(x^k) = (k \cdot l(x)) \pmod{q-1}$$

$$l(x \cdot y) = (l(x) + l(y)) \pmod{q-1}$$

$$l(x/y) = (l(x) - l(y)) \pmod{q-1}$$

$$l(x+y) = l(y) + s\left[(l(x) - l(y)) \pmod{q-1}\right] \pmod{q-1}$$

For the last equation, compare

$$x + y = y(x/y + \mathbf{1}).$$

If $x = \mathbf{0}$ or $y = \mathbf{0}$ we easily calculate the correct result as a special case. We must also treat specially the case that

$$s[l(x) - l(y) \pmod{q-1}] = q - 1 = l(\mathbf{0})$$

which arises when $y = -x$, but this is also easy.

Observe that all of this consists of straightforward local arithmetic except for the single table lookup needed for addition, which is implemented with a 'get' operation. Note too that only when computing powers does the calculation of remainders $\pmod{q-1}$ really require a division (which is relatively slow), the other cases need only a test followed perhaps by addition or subtraction of $q-1$. If subtraction in GF(q) were required we could easily incorporate a "predecessor table" analogous to s.

It is clear now why we avoid the more usual approach. Table lookup will certainly be a significant, and possibly dominant, component of the running time. In typical calculations (*e.g.*, Horner's rule evaluation of dense polynomials) the usual approach requires lookups for both logarithms and antilogarithms; the extra lookup is much more time-consuming the the arithmetic required to compute sums using the successor table.

4.4 Constructing the Successor Table

We continue to work in GF(q), where now we suppose $q = p^n$, p prime. We also take $\pi(x)$ to be a primitive polynomial over GF(p) for GF(q) with $\pi(b) = \mathbf{0}$ for some $b \in$ GF(q). As before, we suppose that our machine has at least q (virtual) processors and let processor i correspond to ι^i; processor $q-1$ corresponds to $\mathbf{0}$. The algorithm which constructs the successor table is given informally as:

1. Calculate $f_i(x) = x^i \bmod \pi(x)$ for all i in parallel and let $a_0[i]$ be the constant term of f_i. $f_{q-1}(x)$ is the zero polynomial.
   - We use the standard fast exponentiation algorithm (repeated squaring), which requires $\log_2 q$ iterations.
   - The calculation involves straightforward polynomial manipulation with coefficients in GF(p), degrees bounded by $2n$.
   - All computation is local arithmetic with small integers.
2. Calculate $r[i] = f_i(p)$ (integer arithmetic) for all i in parallel. $r[i]$ is the position of f_i in the natural lexicographic sequence (the zero polynomial at position 0).
3. Calculate
$$h[i] = \begin{cases} r[i] + 1 & \text{if } a_0[i] < p-1 \\ r[i] - (p-1) & \text{if } a_0[i] = p-1 \end{cases}$$
4. Send i to $z[r[i]]$ for all i in parallel
5. Get $s[i]$ from $z[h[i]]$ for all i in parallel

Observe that $r[i]$ can also be viewed as the value of the radix-p numeral whose digits are the coefficients of f_i, consequently $0 \le r[i] < p^n = q$ for all i. The point of step 3 is that if $h[i] = r[k] = l$ then $f_k(x) = \mathbf{1} + f_i(x)$. Steps 4 and 5 implement a "rendezvous" (in space, not time)[2] so that processor i can identify processor k, the two exchanging information at processor l, whose address can be calculated by both (see figure 4.2).

We should mention that the programs used thus far in our study are designed for fields of arbitrary characteristic and are, in consequence, relatively inefficient when calculating $x^i \bmod \pi(x)$ for fields of characteristic 2. A specialized, and significantly faster, program could be written for this case, but we have not so far felt pressed to do so.

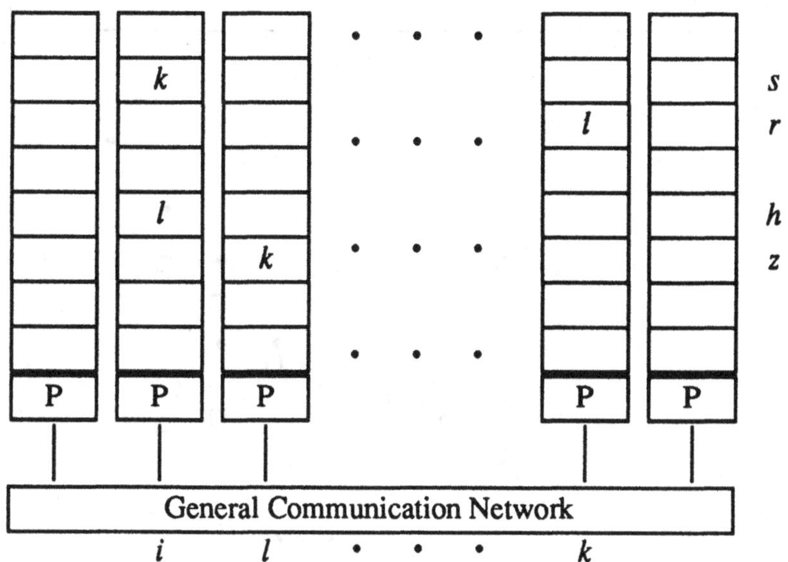

Figure 4.2. Rendezvous to form the Successor Table

4.5 An Example

In studying certain binary cyclic codes, Janwa [3] found it useful to determine the number of rational points of

$$X^{11} + Y^{11} + Z^{11} + (X + Y + Z)^{11}$$

over GF(2^{s}) for $s \leq 21$. We can reduce this to finding the number of roots of

$$X^{11} + Y^{11} + \mathbf{1} = (X + Y + \mathbf{1})^{11} \tag{4.1}$$

We could view this as the problem of counting the total number of roots of a family of monovariate polynomials in X indexed by $Y \in \mathrm{GF}(2^s)$, and the calculation would be entirely feasible for moderate s, but there are important symmetries which can usefully be exploited.

The squaring operator is a bijection on GF(2^s) as, more generally, is the function that maps x to x^p on GF(p^n). Squaring boths sides of (4.1) and using the fact that $(x + y)^2 = x^2 + y^2$ in a field of characteristic 2 we obtain

$$(X^2)^{11} + (Y^2)^{11} + \mathbf{1} = ((X^2) + (Y^2) + \mathbf{1})^{11}$$

If r values of X solve (4.1) for a given Y, then there are exactly r values X' which solve (4.1) for Y^2, and these are just the squares of the solutions for Y. If we determine one representative for each orbit of the squaring operator, we need only

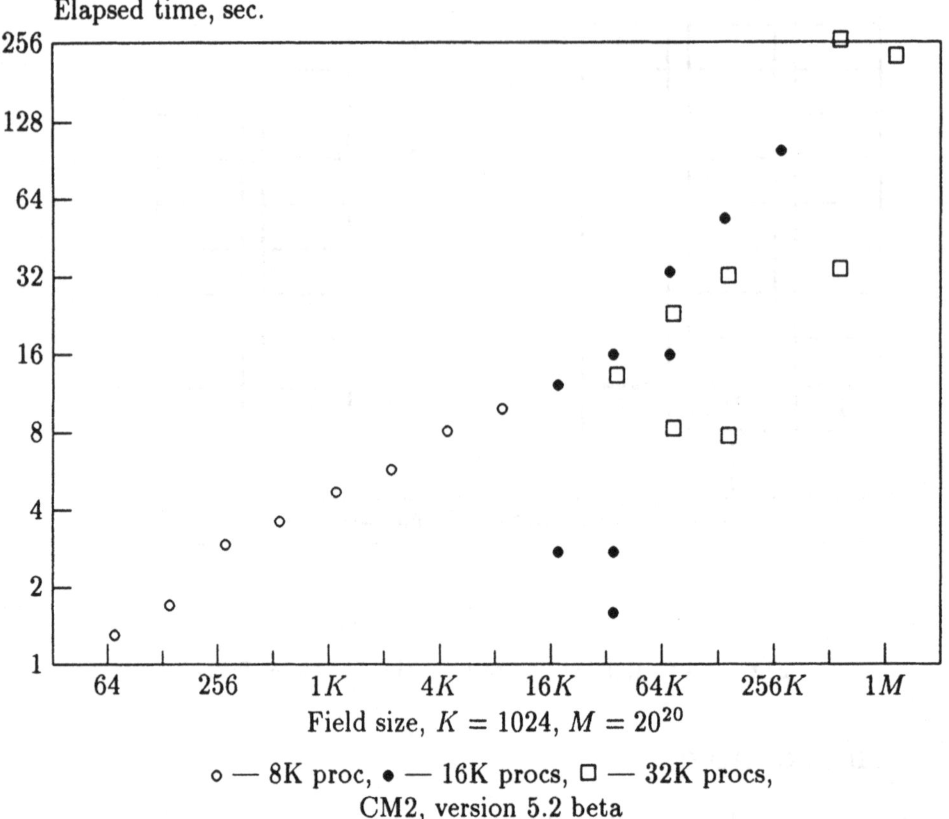

Figure 4.3. Field setup time, including *p*-orbits

solve (4.1) for each representative, multiplying the number of roots by the size of the orbit.

Determining the orbits and selecting representatives is an easy parallel computation; we make use of the fact that no orbit has more than s elements, though some have fewer. Informally, the algorithm is as follows:

1. Each processor (except the one for **0**) calculates its orbit, counting the number of elements.
   - All calculation is local arithmetic with discrete logarithms.
   - We perform s iterations for $GF(2^s)$.
2. A processor identifies itself as a representative if it has the smallest logarithm in the orbit.
3. **0** is the only element of its orbit.

This calculation is performed once, along with construction of the successor table. It can be shown that $GF(2^s)$ has only slightly more than $2^s/s$ orbits, so the

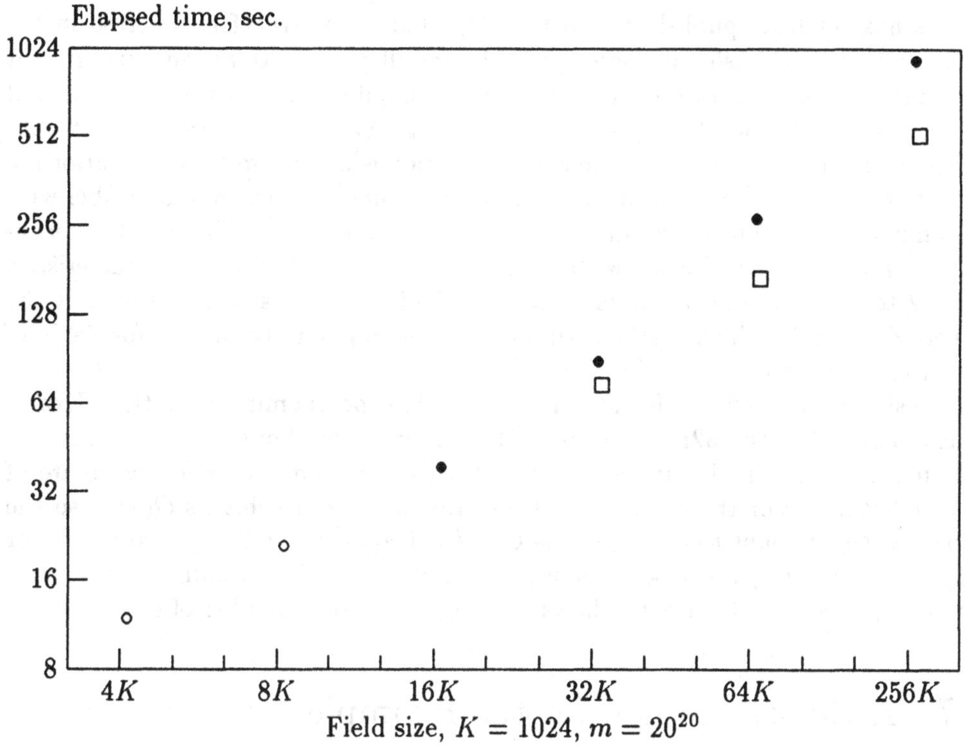

Elapsed time, sec.

1024 512 256 128 64 32 16 8

4K 8K 16K 32K 64K 256K

Field size, $K = 1024$, $m = 20^{20}$

1 ○ — 8K proc, • — 16K procs, □ — 32K procs,
CM2, version 5.2 beta

Figure 4.4. Polynomial solution time, using orbits

effort required is reduced by essentially a factor of s, compared to the exhaustive approach.

4.6 Performance

In estimating performance of the programs we can either take the view that the number of processors available is $O(q)$, which is unrealistic for truly asymptotic estimates but indicative of performance on smaller problems, or we can take the view that the number of processors is fixed, which is appropriate for larger calculations. We take it that the time for operations in GF(p) (for prime p) is $O(\log_2 p)$ which is perhaps optimistic, but not overwhelmingly so. With this assumption, the time required to form the successor table and determine orbits of the squaring (or p-th power) operator is easily seen to be $O(n^2 \log_2 q)$ using $O(q)$ processors, or $O(qn^2 \log_2 q)$ using $O(1)$ processors, if we ignore communication time.

Accounting for communication time is problematic here. The CM router algorithms are proprietary and no detailed analysis of general communication performance has yet been published. On the other hand, the transfers involved in the rendezvous to establish the successor table are all permutations, and experiment shows that time needed for communication is negligible compared to the time used in calculating $x^i \bmod \pi(x)$ by repeated squaring. We know, moreover, that published algorithms [6] for hypercube communication will perform these operations in time $O((\log_2 q)^2)$ with $O(q)$ processors, so the bounds are entirely plausible with communication included. Figure 4.3 shows setup times (including orbit determination) for a number of fields with sizes up to $5^9 = 1,953,125$ and characteristics from 2 to 31. The upper points are for fields of characteristic 2 and exhibit the expected smooth behavior; the seemingly irregular points below are for fields of larger characteristic.

The situation is quite different when we consider polynomial evaluation. Experiment shows that the 32K-processor CM2 can evaluate about 4×10^6 terms/sec. Communication time dominates, so that details of arithmetic performance are of little relevance. For the example problem, the number of orbits is $O(q/n)$ so the time needed to count roots of (4.1) is $O((q/n)(\log_2 q)^2)$ using $O(q)$ processors, or $O(q^2/n)$ using $O(1)$ processors, taking a plausible bound on communication time for $O(q)$ processors. Figure 4.4 shows running times for a number of cases.

4.7 A Better Method for the Example

Janwa and R.M. Wilson [4] have recently shown that the number of rational points of

$$X^{11} + Y^{11} + Z^{11} + (X + Y + Z)^{11}$$

can in fact be determined from the quantity

$$N_s = 3M - 2^s + \sum_{c \in \mathrm{GF}(2^s)} (freq(c))^2$$

where $freq(c)$ is the number of roots of

$$X^{11} + (X + \mathbf{1})^{11} = c$$

and M is the number of roots of

$$X^{10} = (X + \mathbf{1})^{10}$$

with all of these taken over $\mathrm{GF}(2^s)$. Although this approach is quite specialized, it is of interest to see just how easy the parallel computation becomes. The calculation is:

1. Test, in parallel, whether $10i \equiv s[i] \bmod q-1$ for $i < q-1$ (excludes $\mathbf{0}$). Count the number of successes using the (fast) global sum operation to obtain M.
2. Calculate $c[i] = l(X^{11} + (X + \mathbf{1})^{11})$ where $X = b^i$ and set $f[i]$ to 0, for all i in parallel.

3. Send 1 to $f[c[i]]$ with addition, for all i in parallel. This leaves $f[i] = freq(i)$.
4. Set $f[i] := f[i]^2$ for all i in parallel, then form $\sum_{c \in \mathrm{GF}(2^s)} f[i]$ (another global sum) and calculate N_s.

Notice that the entire calculation requires some local arithmetic, one lookup in the successor table (step 2), one 'send-with-add' (also general communication), and two global sums, which are quite fast. For any case of interest, the computation takes no more than about five seconds, far less than the time needed to construct the successor table. Here the CM communication is very effective indeed.

4.8 An Alternative Approach for Small Fields

The method discussed thus far is reasonable for fields with 2^{13} or more elements, and can be adapted to smaller fields by the simple device of running enough copies of the field representation to fill up the hardware being used and dividing the work among these copies. We may, however, wish to explore very large families of polynomials defined over quite small fields, and for these another approach, which does away with interprocessor communication, suggests itself. We can think of each processor as having a copy of the complete successor table in its local memory, as well as the coefficients of one polynomial of the family. Our basic loop steps through the elements of the field, evaluating all polynomials in parallel and identifying the roots.

While we have eliminated interprocessor communication, there is a price to pay. The table lookups now require "indirect addressing" of the processor memories, that is, the processors calculate individual addresses to be used for local memory reference, rather than sharing a common memory address, as is the usual case. Indirect memory reference is, in fact, quite expensive. Experiments conducted on $GF(64)$ produced calculation rates of about 4×10^6 terms/sec. (32K CM-2), essentially the same as with the earlier technique. It is possible that still smaller fields would show some benefit, and also possible that further refinement of the code would help. Table entries were stored as 32-bit values (though 6 would suffice) so that the programs could take advantage of special facilities for indirect access to arrays of 32-bit values. The system also allows such arrays to be shared among groups of 32 processors, at a great saving in memory and perhaps some savings in time. We see no prospects for dramatic speedup from such devices, however.

4.9 Conclusion

We consider that these programs make a useful beginning towards parallel finite-field calculations, though 4×10^6 terms/sec. is not an astounding rate for a 32K CM-2. On the other hand, the programs which construct the successor table yield rates in the neighborhood of 200×10^6 operations/sec. in $GF(p)$, which is quite satisfactory. We note too that the communication time, which dominates polynomial evaluation, degrades only in proportion to field size and not by some larger factor, as can sometimes happen with CM communication using very large vp ratios. Using orbits of the p-th power function to remove symmetries in polynomial families seems to be a convenient technique of fairly broad application, though as we have seen, cleverly devised transformations may yield enormous reductions in the computational effort in particular instances.

For the future we plan to implement specialized code for constructing the successor table in fields of characteristic 2, code which should be several times faster than the present general-purpose programs. Using this code, we shall also explore a mixture of table lookup with direct polynomial multiplication (mod $\pi(x)$) and polynomial (*i.e.*, vector) addition. Such a strategy might well prove effective for small fields of characteristic 2, and just possibly for fields of larger characteristic.

Acknowledgements. We thank P. Solé for many helpful suggestions, H. Janwa and J. Wolfmann for suggesting problems of interest, and H-M Chen, N. Copty, and Y. Kung for programming contributions. Computing resources were generously provided by the Northeast Parallel Architectures Center (NPAC) at Syracuse University.

4.10 References

[1] Hillis, W.D. (1985). *The Connection Machine.* The MIT Press, Cambridge, MA.

[2] Hillis, W.D. and Steele, G.L., Jr. (1986). Data Parallel Algorithms. *Commun. ACM* 29, 12, 1170-1183.

[3] Janwa, H. (1990, March). Private communication.

[4] Janwa, H. and Wilson, R.M. (1990, May). Hyperplane Sections of Fermat Varieties in P^3 in Positive Characteristic and Some Applications. Preprint.

[5] Lidl, L., and Niederreiter, H. (1986). *Introduction to finite fields and their applications.* Cambridge University Press, Cambridge, UK.

[6] Nassimi, D. and Sahni, S. (1981). Data Broadcasting in SIMD Computers. *IEEE Trans. Comp.*, C-30, 2, 101-107.

[7] Thinking Machines Corporation. (1988, September). *Lisp Reference Manual, Version 5.0. Cambridge, MA.

[8] Thinking Machines Corporation. (1989, May). Connection Machine Model CM-2 Technical Summary, Version 5.1, Cambridge, MA.

5

Embarrassingly Parallel Algorithms for Algebraic Number Arithmetic — and some less trivial issues

Dennis Weeks[1]

Abstract: Representing algebraic numbers α and β by their defining polynomials is an alternative to the older representation in which the sum $\alpha + \beta$ would be represented by a list of character strings recursively involving root indices, $+$, $-$, $\times$, $/$, and integer terms or radicands. Algorithms for sums, products, etc., in the defining polynomial representation, are based on the theory of symmetric functions and have relatively efficient implementations using polynomial resultants. Even better algorithms use power sums rather than the coefficients of the defining polynomials; on massively parallel systems $\alpha \times \beta$ executes in constant time, and computation of $\alpha + \beta$ is linear (or logarithmic, if enough processors are available) in mn, where m and n are the degrees of the defining polynomials and mn is the degree of the result. However, given polynomials of degree m and n, these algorithms require their power sums to order mn. The best known power-sum algorithm is based on Newton's identity, which may be treated as a linear recurrence, whose solution, conventionally understood to be of complexity $O(mn)$, will dominate the time of the multiplication algorithm and significantly increase the addition time. A parallel algorithm, reducing the recurrence solution to $O((\log_2 n)(\log_2 m))$, is discussed.

Keywords: algebraic numbers, power sums, linear recurrences

5.1 Introduction

The basic concept of the defining polynomial is quite straightforward: if P is a polynomial, and α is a root of the equation $P = 0$, then P "defines" α. In computer implementations obviously the coefficients of P should be integers[1], or pairs of integers representing rational numbers. We impose two further restrictions on P:

(1) P is irreducible over the coefficient domain;

[1]MasPar Computer Corporation, 749 North Mary Ave., Sunnyvale, CA 94086

[1]The coefficients of P could be in any integral domain; for brevity we do not discuss here the computational or representational issues that could arise, say, if the coefficients were in a field of characteristic $\neq 0$.

(2) P is primitive (GCD of its coefficients is 1) or monic (its leading coefficient is 1).

These restrictions guarantee uniqueness. The choice of "primitive" vs. "monic" essentially reflects whether the coefficients are integers or rational numbers: if the coefficients are integers, primitivity is a sufficient condition, while if rational numbers are used then the polynomials must all be monic. (A primitive polynomial is in fact also monic if and only if all of its roots are *algebraic integers.*)

For convenience we will write $P = \overline{\pi}(\alpha)$ to signify that P, a polynomial, is the defining polynomial of α, an algebraic number; if $\alpha = \overline{\rho}(P)$ (i.e., α is a root of $P = 0$) but P fails to satisfy the irreducibility or primitivity criteria, then we may write $P \supset \overline{\pi}(\alpha)$ where some factor of P, say Q, is the proper defining polynomial of α.

Representation of an algebraic number by its defining polynomial in fact requires more than just a list of the polynomial's coefficients: if the polynomial is of degree n, it has n distinct roots (if it satisfies the irreducibility criterion, a defining polynomial is squarefree, i.e. it has no multiple roots), and so to complete the representation of an algebraic number we need another integer, indicating which one of the defining polynomial's roots is the one of interest, or a pair of floating-point or rational numbers defining an interval containing the root of interest and no other root (two such pairs if the root has real and imaginary parts). The choice of ordering schemes, and methods of determining the location of the root in question among all the roots of the polynomial, are non-trivial issues which however are beyond the scope of this paper.

5.2 Algebraic Number Arithmetic: The Basic Algorithms

The root product algorithm: Let $P = \overline{\pi}(\alpha)$ and $Q = \overline{\pi}(\beta)$ be input polynomials of degrees i and j respectively. Then $R = \overline{\pi}(\alpha\beta)$ will be a polynomial of degree $k = ij$ whose roots include $\alpha\beta$ and all other products of (one root of P)×(one root of Q). Now let p_n represent the sum of the nth powers of the roots of P, and likewise q_n and r_n are the corresponding power sums of the roots of Q and R. Then

$$r_n = p_n q_n$$

for all non-negative integer values of n. This algorithm appeared in the "gray literature" (Selmer[1966*a*, p. 91], by whose notation we are computing $R = P\S Q$) and was independently derived by Weeks[1974] based on a model of Hirsch[1827] for the power sums of the polynomial whose roots are the sum of all pairs of roots of a single given polynomial.

The root sum algorithm: With the same definitions as above, let $S \supset \overline{\pi}(\alpha + \beta)$ (the irreducibility criterion for the defining polynomial is not necessary for the validity of either the sum or product algorithm) and let s_n be the nth power sum of the roots of S. Then

$$s_n = \sum_{i=0}^{n} \binom{n}{i} p_i q_{n-i} \quad \text{[Weeks 1974].}$$

Now let us introduce an analogy to the $\overline{\pi}$ notation: if $P = \overline{\pi}(\alpha)$ is of degree n, and T is a list (vector) of the first $m \geq n$ power sums of the roots of P, then we say $T = \overline{\sigma}(\alpha)$, and for completeness conversely $\alpha = \overline{\tau}(T)$. We can concatenate this with the previous notation, whereby "$T = \overline{\sigma}(\overline{\rho}(P))$" asserts that T is a list of the power sums of the roots of P. I find this notation convenient, because for instance the root product algorithm can be written in this notation as $\overline{\sigma}(\alpha\beta) = \overline{\sigma}(\alpha) \cdot \overline{\sigma}(\beta)$ (where in this case multiplication is taken to be formation of products of corresponding elements of the two input vectors.)

Other interesting algorithms. The purpose of this paper is to discuss arithmetic operations on algebraic numbers, but some other polynomial operations also can be done extremely easily in the power sum domain, and deserve passing mention here:

(1) Let U and V be two polynomials, let A be the set of all roots of U, and let B be the set of all roots of V. If $W = UV$ (conventional polynomial multiplication), and C is the set of all roots of W, then obviously $C = A \cup B$ (strictly speaking, C may be a "multiset" because A and B may have elements in common or may be multisets themselves, and in either case UV could have multiple roots). Then the power sums of W are constructed by adding the corresponding power sums of U and V, i.e. $\overline{\sigma}(\overline{\rho}(W)) = \overline{\sigma}(\overline{\rho}(U)) + \overline{\sigma}(\overline{\rho}(V))$, where "+" is taken to be componentwise addition. The algorithm for the product of two polynomials, represented by their power sums, is thus very much like the algorithm for the product of their roots: we have simply substituted power sum addition in place of power sum multiplication.

(2) A classic representation of the discriminant of a polynomial is the determinant

$$D = \begin{vmatrix} s_1 & s_2 & \cdots & s_{n-1} & s_n \\ s_2 & s_3 & \cdots & s_n & s_{n+1} \\ & & \vdots & & \\ s_{n-1} & s_n & \cdots & s_{2n-3} & s_{2n-2} \\ s_n & s_{n+1} & \cdots & s_{2n-2} & s_{2n-1} \end{vmatrix}$$

where the elements are the power sums of the roots of the polynomial. Also, the Borchardt-Jacobi theorem determines the number of real roots of the polynomial, by computing the "signature" of the discriminant matrix, i.e. the number of changes of sign in the sequence

$$M_1,\ M_2,\ \ldots,\ M_{n-1},\ D$$

where the M's are determinants of the principal minors of the discriminant matrix (MacDuffee[1931] refers to the signature as the difference between the numbers of positive and negative terms in the sequence; this disagrees with Borchardt's[1847] original statement of the theorem).

5.3 Parallel Implementation

Obviously the hearts of the sum and product algorithms are ideal candidates for parallel implementation: each power sum of the result is generated independently of the others. In the product algorithm, in fact, only one arithmetic operation, a multiplication, is required, so with a sufficient number of processors it executes in constant time regardless of the degree of the result. The sum algorithm requires $2(n+1)$ multiplications and n additions for the nth power sum of the result, so it is $O(k)$ for an output of degree k, if at least k processors are available. (If $k(k+1)/2$ processors are available, the sum algorithm can be reduced to $O(\log_2 k)$ by a "divide and conquer" technique that is commonly used for operations such as inner products and "sum reductions" on massively parallel machines, although the cost of communication and replication of data across mutiple processors may somewhat counteract the benefits of the more clever summation algorithm.)

This is the "embarrassingly parallel" aspect of the algorithms: doing this part in parallel is so trivial it's hardly worth mentioning. However, there are other things we like to do to polynomials, where an algorithm based on power sums is extremely difficult, or does not exist. "Extending the basis", i.e. producing power sums of order mn from a minimal set of the first n power sums, is an issue which will arise virtually every time we want to add or multiply two algebraic numbers in the defining polynomial representation. It is possible but very difficult, unless the coefficients are also involved in the computation. Although the discriminant will tell us if a result has multiple roots, the more general issues of verifying irreducibility, and of separating the result into irreducible factors, require work in the coefficient domain (perhaps only because algorithms based on power sums have not been invented).

Therefore, despite the convenience of power sums for some problems, we have to be able to switch between the power sum and coefficient domains, and the rest of this paper will be concerned with how that can be done on a parallel system.

5.4 Newton's Identities in Data-Parallel Computers

A number of algorithms, both classical and modern, exist for the production of power sums from coefficients, and coefficients from power sums. "Waring's formulas" express a power sum of order m as a polynomial involving all coefficients up to the degree of the polynomial or m, whichever is less. With no further information, this would appear to be a logical choice for parallel computation, since each power sum could be computed independently of the others (unlike the Newtonian recurrence, discussed below). However, the number of terms in the polynomial, and the number of coefficients which must be multiplied together to form any particular term, correspond to the number and structure of all of the integer partitions of m, the order of the power sum being constructed; the number of processors has to grow exponentially, and the worst case multiplication on some processor grows linearly with m.

This means, for better or worse, that some variant of the Newtonian algorithm is what we must consider. With c_i representing the coefficient of x^{n-i}, and s_i representing the sum of the ith powers of the roots of the polynomial (the following formulas assume that $c_0 = 1$, i.e. the polynomial is monic, and $s_0 = n$, the degree of the polynomial), the Newtonian recurrence says that

$$
\begin{aligned}
s_1 &= -\ 1 \cdot c_1,\\
s_2 &= -\ 2 \cdot c_2 - s_1 \cdot c_1,\\
s_3 &= -\ 3 \cdot c_3 - s_1 \cdot c_2 - s_2 \cdot c_1,\\
&* * *\\
s_n &= -s_0 \cdot c_n - s_1 \cdot c_{n-1} - \ldots - s_{n-2} \cdot c_2 - s_{n-1} \cdot c_1,\\
s_{n+1} &= -s_1 \cdot c_n - s_2 \cdot c_{n-1} - \ldots - s_{n-1} \cdot c_2 - s_n \cdot c_1,\\
s_{n+2} &= -s_2 \cdot c_n - s_3 \cdot c_{n-1} - \ldots - s_n \cdot c_2 - s_{n+1} \cdot c_1,
\end{aligned}
$$

etc., where for power sums of index greater than n the formula is a classical order-m recurrence, each result depending on the n previous results and the n coefficients of the input polynomial.

A straightforward implementation of this algorithm to produce all the power sums to order N of the roots of a polynomial of degree m, would have an inner loop performing steps of the following type:

```
s[i] = s[i] + s[i-j] * c[j]
```

and any vectorizing compiler would report that it could not emit code to generate, say, `s[k]` and `s[k+1]` in parallel, since `s[k+1]` is recursively dependent on `s[k]`. (A vector machine could perform the m multiplications for `s[k]` in the vector registers, but the computation time would still be $O(m)$ for each step, and $O(m^2)$ or $O(Nm)$ overall.) Similar recurrences produce coefficients from power sums, or the "homogeneous product sums" from either coefficients or power sums.

The best strategy to compute the power sums, in parallel, from the coefficients of a polynomial, appears to be one outlined by Kogge[1974], in which an approach similar to "divide and conquer" for vector sum reduction is used to solve general linear recurrences in parallel in $\log_2 N$ outer loop steps, where N is the length of the recurrence. Before going into the details of Kogge's algorithm, let us formally recast Newton's identities into a linear recurrence problem.

In a linear recurrence of order m, each step can be written

$$s_j = \mathrm{f}(\vec{a}_j, s_{j-1}, s_{j-2}, \ldots, s_{j-m}) \qquad j = 1\,..\,N$$

where $\vec{a}_j$ is the *parameter vector* for step j; in general, each $\vec{a}$ will have m elements, and $\vec{a}_j \neq \vec{a}_{j-1}$. If $j \leq m$, s_i terms where i is negative or zero appear in this formulation; those are the *initial conditions* of the problem.

Applying this specifically to the power sums of the roots of an equation $x^m +$

$c_1x^{m-1} + c_2x^{m-2} + \ldots + c_{m-1}x + c_m = 0$, we have (cf. Selmer[1966b]

$$
\begin{aligned}
s_0 = m,\ s_{-1} &= m-1,\ s_{-2} = m-2,\ \ldots,\ s_{1-m} = 1\,; \\
\vec{a}_1 &= (\ \ 0,\quad 0,\quad 0,\quad \ldots,\ 0,\ -c_1), \\
\vec{a}_2 &= (-c_1,\quad 0,\quad 0,\quad \ldots,\ 0,\ -c_2), \\
\vec{a}_3 &= (-c_1,\ -c_2,\quad 0,\quad \ldots,\ 0,\ -c_3), \\
&\ \ldots \\
\vec{a}_{m-1} &= (-c_1,\ -c_2,\ \ldots,\ -c_{m-3},\ 0,\ -c_{m-1}), \\
\ldots = a_{m+1} = a_m &= (-c_1,\ -c_2,\quad \ldots,\ -c_{m-1},\ -c_m)
\end{aligned}
$$

the function is essentially a dot product of the parameter vector and a vector composed of the m previous s values.

The key to Kogge's paper is the observation that the parameter vectors $\vec{a}_i$ are all independent and known at the beginning of the problem (he is not addressing on-line recursive filtering), therefore with a little algebra we can determine how, say, s_{j+2} can be produced from s_j, s_{j-1}, ... and $\vec{a}_{j+2}$, $\vec{a}_{j+1}$ without explicit involvement of s_{j+1}. With this established, one can then advance from "leapfrogging" alternate s terms, to leapfrogging 3, 7, or more at a time. The strategy is similar in concept to the "rank k updates" that have become increasingly popular among numerical analysts solving large matrix problems on parallel systems. By a progression of steps, each one effectively doubling the span of parallelism, the algorithm eventually reaches a point where every term in the recurrence can be computed in parallel from the initial conditions and a "companion set" of vectors generated during the "recursive doubling" stages. The outer loop is $O(\log_2 N)$, where N is the index of the highest power sum required as output; inner loops may also be done in parallel at $O(\log_2 m)$. The final step is to compute a dot product of the auxiliary parameter vector with the vector of initial conditions, a process which is $O(m)$ in a serial environment, but if enough processors are available all the multiplications can be done simultaneously, and the summation of the individual products using the "divide and conquer" technique is now $O(\log_2 m)$. If $N \gg m$ this latter term will be insignificant. Typically we will use this algorithm in the following way:

(1) Given two polynomials, P of degree m and Q of degree n, to produce a polynomial of degree $N = mn$.

(2) Use the Kogge procedure to produce N power sums from the m coefficients of P.

(3) Use the Kogge procedure to produce N power sums from the n coefficients of Q.

(4) Do the $O(1)$ root-product or $O(\log_2 N)$ root-sum computation.

(5) Use the Kogge procedure to produce N coefficients from the N power sums of the result.

5.5 Summary

The basic algorithms for sums and products of algebraic numbers, using power sums of the roots of the defining polynomials, are easy to parallelize and have parallel complexity of 1 (root products) or log mn (root sums), where m and n are degrees of the input polynomials. These procedures generally need to be bracketed with procedures to convert coefficients to power sums and vice versa, which can also be done with parallel complexity proportional to log mn. To convert between power-sum and coefficient domains by Kogge's procedure may not be the most effective method, as it is very general and does not take into account, for instance, that this latter process is essentially the solution of a triangular matrix problem. An algorithm tuned for this specific recurrence might be able to do better; while it would still probably be $O(\log_2{}^2 N)$, its multiplier might be smaller than in the general case (cf. Sameh & Brent [1977]).

We have also mentioned that the product of two polynomials can be computed in constant time using power sums instead of coefficients. If we count the time needed to produce $m+n$ power sums from the m and n coefficients of the two inputs, and to produce $m+n$ coefficients from the power sums of the result, we have an angorithm of parallel complexity $O(\log_2{}^2(m+n)+(\log_2 m)(\log_2(m+n))+(\log_2 n)(\log_2(m+n)))$, which is competitive with FFT-based polynomial multiplication algorithms (cf Ponder[1989], Siebert-Roch and Muller[1989]). Indeed, the structural similarity between the FFT-based algorithm and polynomial multiplication using power sums makes it tempting to regard power sums as a "Discrete Newton Transform" of coefficients; this view of the process could lead to numerous insights for other polynomial-oriented procedures.

5.6 References

Borchardt, C. W. [1847] Développements sur l'équation à l'aide de laquelle on détermine les inégalités séculaires du mouvement des planètes. *Jour. de Math.* (Liouville) 12:50–67.

Hirsch, M. [1827] *Hirsch's Collection of Examples, Formulae, and Calculations of literal Calculus and Algebra.* Translated from the German, by the Rev. J. A. Ross, A. M. London: Black, Young, & Young (1st German edition, Berlin: Duncker und Humblot, 1804.)

Kogge, P. [1974] Parallel solution of recurrence problems. IBM *Jour. of Res. and Dev.* 18(1974):138–148.

Loos, R. and G. E. Collins [1974] Resultant algorithms for exact arithmetic on algebraic numbers. *SIAM Rev.* Jan. 1974: 130–131.

MacDuffee, C. C. [1931] The discriminant matrices of a linear associative algebra. *Annals of Math., 2nd Ser.* 32(1931): 60–66.

Ponder, C. G. [1989] Evaluation of "Performance Enhancements" in algebraic manipulation systems (thesis abstract). In *Computer Algebra and Parallelism* (Ed. J. Della Dora and J. Fitch), Academic Press, 1989, pp. 51–73.

Selmer, E. S. [1966*a*] *Linear recurrence relations over finite fields.*(mimeograph) Department of Mathematics, Univ. of Bergen, Norway, 212 pp.

Selmer, E. S. [1966*b*] On Newton's equations for the power sums. *BIT* 6(1966): 158–160.

Siebert-Roch, F. and J.-M. Muller [1989] VLSI manipulation of polynomials. In *Computer Algebra and Parallelism* (Ed. J. Della Dora and J. Fitch), Academic Press, 1989, pp. 233–256.

Weeks, D. [1974] Formal representations for algebraic numbers. *SIGSAM Bulletin* #31 (ACM) (August 1974): 91–95.

6

Parallel Real Root Isolation Using the Coefficient Sign Variation Method

G. E. Collins[1]
J. R. Johnson[1]
W. Küchlin[1]

Abstract: We present a parallel implementation of the coefficient sign variation method for polynomial real root isolation. The implementation uses PARSAC-2, a parallel version, based on threads, of the SAC-2 computer algebra system. A discussion of the implementation and its performance is given. Our timing results were obtained on a shared memory multiprocessor implementation using the Encore Multimax.

6.1 Introduction

In this paper we discuss the parallel implementation and performance of the coefficient sign variation method for real root isolation of integral polynomials. A real root isolation algorithm produces isolating intervals (intervals containing exactly one root) for all of the real roots of a polynomial. The coefficient sign variation method uses Descartes' rule of signs to search for isolating intervals. The algorithm checks an initial interval and, if necessary, bisects the interval and recursively searches the left and right subintervals. The algorithm is parallelized by performing the left and right recursive calls in parallel.

The performance of the algorithm is determined by a binary tree associated with the search the algorithm performs. We empirically investigate properties of this tree which affect the computing time and the amount of parallelism that can be obtained with this algorithm.

In Section 6.2 we review the coefficient sign variation method and in Section 6.3 we discuss its computing time. In Section 6.4 we describe our implementation on the Encore Multimax, a shared memory parallel computer. Finally, in Section 6.5 we report empirical results which strongly suggest that the average amount of parallelism is limited. Despite this limitation, the algorithm achieves the predicted

[1]Department of Computer and Information Science The Ohio State University Columbus, OH 43210

speedup, and in cases where more parallelism is available the parallel performance improves accordingly. Moreover, these empirical results lead to a conjectured average computing time for the sequential algorithm which is much better than the worst case bound in Section 6.3.

6.2 Review of the Sign Variation Method

In this section we review and modify the coefficient sign variation method for polynomial real root isolation proposed by Akritas and Collins in [1] and reviewed by Collins and Loos in [6]. The algorithm will be modified slightly to simplify the parallel program. The only difference from the original algorithm is that the positive and negative roots are isolated simultaneously. With this change, the extra parallelism available from the positive and negative roots is obtained using the recursive parallel calls of the algorithm.

The coefficient sign variation method is based on two special cases of Descartes' rule of signs. Descartes' rule relates the number of positive roots of a polynomial to the number of coefficient sign variations of the polynomial. Let $A(x) = \sum_{i=1}^{n} a_i x^{e_i}$ be a polynomial with real coefficients, where $a_i \neq 0$ and $e_1 < e_2 < \cdots < e_n$. The number of variations of $A(x)$, $\mathrm{var}(A(x))$, is equal to the number of values of i such that $a_i \cdot a_{i+1} < 0$. For example, if $A(x) = 3x^7 - 5x^4 + 3x^3 + 2x^2 - x + 7$ then $\mathrm{var}(A(x)) = 4$. Descartes' rule states that the number of variations of a polynomial exceeds the number of positive roots by a non-negative even number. This implies that, if $\mathrm{var}(A(x)) = 0$ then $A(x)$ has no positive roots, and if $\mathrm{var}(A(x)) = 1$ then $A(x)$ has one positive root. These two special cases of Descartes' rule are used to search for isolating intervals.

The coefficient sign variation method begins by computing a root bound B for $A(x)$ (See [13] Sec. 4.6.2 Ex. 20 for the root bound we use). Then the interval $(-B, B)$ is repeatedly bisected until isolating intervals have been found for all of the real roots. An interval is checked to see whether it is an isolating interval by applying Descartes' rule. In order to apply Descartes' rule the polynomial $A(x)$ must be transformed to a polynomial $\tilde{A}(x)$ whose positive roots correspond to the roots of $A(x)$ in the interval in question. The variations of the transformed polynomial are then counted. There are three possible outcomes. First, if $\mathrm{var}(\tilde{A}(x)) = 0$ then there are no roots in the interval and the interval can be discarded. Second, if $\mathrm{var}(\tilde{A}(x)) = 1$ then an isolating interval has been found. Finally, if $\mathrm{var}(\tilde{A}(x)) > 1$ then the outcome is undecided and the interval must be bisected.

The algorithm **IPRISS** uses the coefficient sign variation method as outlined above. To ensure termination, **IPRISS** assumes that its input is a squarefree polynomial. To obtain a precise formulation of **IPRISS**, we need to discuss the polynomial transformations that are used. For simplicity, the input polynomial is transformed to a polynomial with all of its roots (real and complex) in the unit circle. This is done by scaling by the root bound. If B is a root bound, then the roots of $A(Bx)$ in the unit circle correspond to the roots of $A(x)$. After this scaling operation, **IPRISS** uses the subalgorithm **IPRISMU** to find isolating intervals contained in the interval $(-1, 1)$.

IPRISMU first checks to see whether the right interval $(0, 1)$ is an isolating interval. This is done by counting the number of variations of $\tilde{A}_2(x) = (x+1)^m A(1/(x+1))$ ($m = \deg(A(x))$). The roots of $\tilde{A}_2(x)$ in the right half plane correspond to the roots of $A(x)$ in the circle of radius $1/2$ centered at the point $(1/2, 0)$ in the complex plane. If more than one variation is obtained, then the polynomial is bisected and the algorithm is called recursively. The right bisection polynomial $A_2^*(x) = 2^m A((x+1)/2)$ is the polynomial whose roots in the unit circle correspond to the roots of $A(x)$ in the circle of radius $1/2$ centered at $(1/2, 0)$. After checking the right subinterval, **IPRISMU** checks to see whether the midpoint is a root. Finally, the left interval $(-1, 0)$ is checked by counting the variations of $\tilde{A}_1(x)$, where $\tilde{A}_1(x) = (-x-1)^m A(1/(-x-1))$. The roots of $\tilde{A}_1(x)$ in the right half plane correspond to those of $A(x)$ in the circle of radius $1/2$ centered at $(-1/2, 0)$. If necessary, the left interval is bisected and the algorithm is called recursively with $A_1^*(x) = 2^m A((x-1)/2)$. Figure 6.1 contains listings of **IPRISS** and **IPRISMU**.

The computation performed by **IPRISS** on

$$\begin{aligned} A(x) = {} & -369x^{20} + 979x^{19} + 190x^{18} + 677x^{17} - 926x^{16} + 179x^{15} - \\ & 91x^{14} - 62x^{13} + 902x^{12} + 980x^{11} + 914x^{10} - 993x^{9} + 3x^{8} - \\ & 505x^{7} + 415x^{6} - 916x^{5} + 592x^{4} - 243x^{3} - 788x^{2} - 81x + 127 \end{aligned}$$

is shown in Figure 6.2. The intervals checked by the algorithm correspond to the horizontal diameters of the circles. The solid circles indicate that an isolating interval was found.

The correctness of the algorithm depends on the correctness of the polynomial transformations and the special cases of Descartes' rule. However, Descartes' rule does not guarantee the termination of the algorithm. To prove termination, we need the following two theorems from [5].

Theorem 1 *If $A(x)$ does not have any real or complex roots in the circle C of radius $\frac{1}{2}$ centered at $(\frac{1}{2}, 0)$, then* $\mathrm{var}((x+1)^m A(1/(x+1))) = 0$.

Theorem 2 *If $A(x)$ has a single root in the circle C and all of the remaining real and complex roots are outside the circles C_1 and C_2 of radius 1 centered at* $(0,0)$ *and* $(1,0)$, *then* $\mathrm{var}((x+1)^m A(1/(x+1))) = 1$.

Since each bisection spreads the roots by a factor of two, eventually the hypotheses of one of these theorems will be satisfied.

6.3 Computing Time Analysis

In this section we present a model which can be used to describe the computation performed by the coefficient sign variation method. This model is used to obtain a worst case bound on the computing time of the algorithm. Our computing time analysis will use notation and concepts introduced by Collins [3]. In the following

$L \leftarrow$ **IPRISS**(A)

[Integral polynomial root isolation, symmetric coefficient sign variation method. A is a squarefree integral polynomial. L is a list of isolating intervals for A.]

1. [Compute root bound of $A(x)$.] $B \leftarrow$ **IUPRB**(A).
2. [Transform roots to the unit circle.] $\overline{A}(x) \leftarrow A(Bx)$.
3. [Isolate roots.] **IPRISMU**$(\overline{A}, -1, 1)$.
4. [Scale intervals.] replace intervals (a, b) in L by (Ba, Bb) ∎

$L \leftarrow$ **IPRISMU**(A, a, b)

[Integral polynomial root isolation, symmetric coefficient sign variation method, unit circle. A is a squarefree integral polynomial. a and b are rational numbers with $a < b$. Let T be a rational linear fractional transformation which maps the interval $[-1, 1]$ to $[a, b]$. Let $T(A)$ be an integral polynomial whose roots in $[a, b]$ correspond to the roots of A in $[-1, 1]$. L is a list of one point or open isolating intervals for the roots of $T(A)$ in the interval (a, b).]

1. [Initialize.] $L_1 \leftarrow ()$; $L_0 \leftarrow ()$; $L_2 \leftarrow ()$;
   if $x|A(x)$ then $\overline{A}(x) \leftarrow A(x)/x$ else $\overline{A}(x) \leftarrow A(x)$; $m \leftarrow \deg(\overline{A})$.
2. [Test right.] $\tilde{A}(x) \leftarrow x^m\overline{A}(1/x)$; $\tilde{A}_2(x) \leftarrow \tilde{A}(x+1)$;
   if $\mathrm{var}(\tilde{A}_2(x)) = 1$ then $L_2 \leftarrow ((\frac{a+b}{2}, b))$.
3. [Right recursive call.] if $\mathrm{var}(\tilde{A}_2(x)) > 1$ then
   $\{\ A^*(x) \leftarrow 2^m\overline{A}(x/2)$; $A_2^*(x) \leftarrow A^*(x+1)$;
   $L_2 \leftarrow$ **IPRISMU**$(A_2^*, \frac{a+b}{2}, b)\ \}$.
4. [Test midpoint.] if $\deg(\overline{A}(x)) \neq \deg(A(x))$ then $L_0 \leftarrow ([\frac{a+b}{2}, \frac{a+b}{2}])$.
5. [Test left.] $\tilde{A}_{11}(x) \leftarrow \tilde{A}(-x)$; $\tilde{A}_1(x) \leftarrow \tilde{A}_{11}(x+1)$;
   if $\mathrm{var}(\tilde{A}_1(x)) = 1$ then $L_1 \leftarrow ((a, \frac{a+b}{2}))$.
6. [Left recursive call.] if $\mathrm{var}(\tilde{A}_1(x)) > 1$ then
   $\{\ A_1^*(x) \leftarrow A^*(x-1)$; $L_1 \leftarrow$ **IPRISMU**$(A_1^*, \frac{a+b}{2}, b)\ \}$.
7. [Concatenate lists of intervals.] $L \leftarrow \mathrm{concat}(L_1, L_0, L_2)$ ∎

Figure 6.1. Algorithms **IPRISS** and **IPRISMU**.

sections, the model will be used to describe and empirically analyze the amount of parallelism in the algorithm and the average behavior of the algorithm.

A trace of the algorithm **IPRISMU** can be represented by a binary tree, where each node in the tree corresponds to a recursive call of the algorithm. For example, Figure 6.3 traces the computation performed on the example given in Section 6.2.

The computing time of the algorithm can be estimated from the tree. For each node some polynomial transformations are performed. The cost of these transformations dominate the cost of the work associated with the node. Four translations ($A'(x) = A(x \pm 1)$), one inversion ($A'(x) = x^m A(1/x)$), one negation ($A'(x) = A(-x)$), and one homothetic transformation ($A'(x) = 2^m A(x/2)$) are needed for an internal node with a right and left recursive call. Three translations, one inversion, one negation, and one homothetic transformation are needed for an internal node with a single recursive call. Two translations, one inversion, and one negation are needed for any leaf node.

Since the size of the coefficients of the input polynomial vary from node to node, the cost of the transformations at a node depends on the node. A bound on the coefficient size at a particular level can be obtained.

Theorem 3 (Coefficient bound.) *Let* $\deg(A(x)) = m$ *and* $|A(x)|_\infty = d$ *(the max norm of* $A(x)$*, which is equal to the maximum of the absolute values of the coefficients [3]). Then the max norm of any polynomial at level* l *is less than or equal to* $2^{2ml}d$.

PROOF. The proof is by induction on l using $|A(x \pm 1)|_\infty \leq 2^m |A(x)|_\infty$ and $|2^m A(x/2)|_\infty \leq 2^m |A(x)|_\infty$ ∎

Using this coefficient bound, we can bound the cost of any node. The cost of all of the transformations are dominated by the cost of a translation. The cost of a translation is given by the following theorem.

Theorem 4 (Computing Time of Translation) *Let* $\deg(A(x)) = m$ *and* $|A(x)|_\infty = d$*. Then, using Horner's method,* $A(x \pm 1)$ *can be computed in time dominated by* $m^3 + m^2 L(d)$*, where* $L(d)$ *is the length of* d.

PROOF. By Theorem 3 the largest coefficient has length $m + L(d)$. Since Horner's method performs $m(m-1)/2$ coefficient additions or subtractions on integers of at most this size, the computing time is dominated by $m^2(m + L(d))$ ∎

These observations along with Theorems 1 and 2 in Section 6.2 can be used to obtain a bound on the computing time of the coefficient sign variation method.

Theorem 5 *Let* $\deg(A(x)) = m$ *and* $|A(x)|_\infty = d$*. Then the computing time of* **IPRISS** *is dominated by* $m^6 \mathrm{L}(md)^2$.

PROOF. The cost of **IPRISS** is determined by the number of nodes in the binary search tree associated with **IPRISMU**, and the cost of each node. This can be estimated by summing the cost of the nodes at each level of the tree. After the initial scaling by the root bound, which is dominated by d, the max norm of the input to **IPRISMU** is dominated by d^{m+1}. The time for **IPRISMU** is dominated

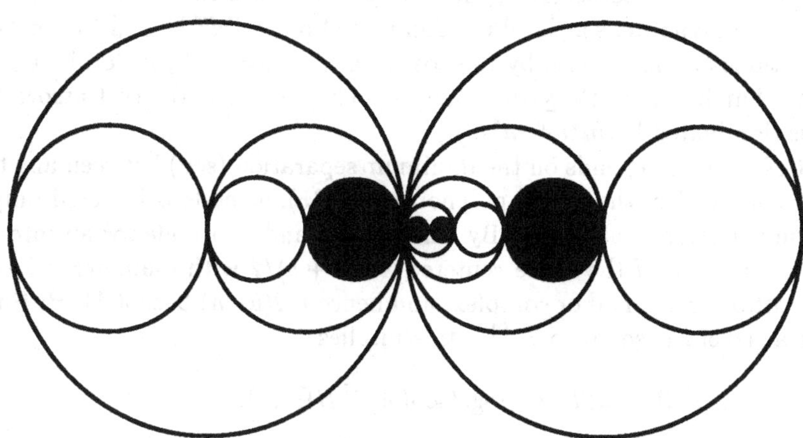

Figure 6.2. Bisections performed by **IPRISMU**.

v1=3
v2=5

v1=0
v2=3

v1=5
v2=0

v1=0
v2=1

v1=2
v2=1

v1=2
v2=0

v1=1
v2=1

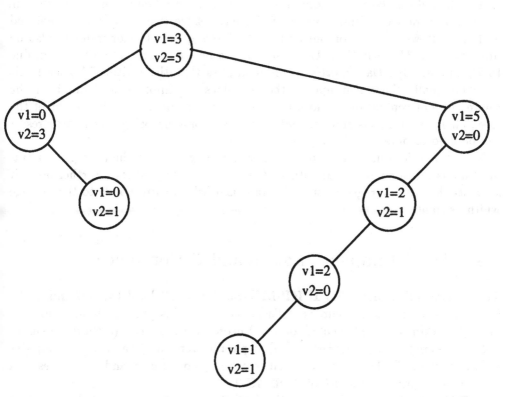

Figure 6.3. Binary tree representing the trace of **IPRISMU**.

by $\sum_{i=0}^{H} W_i C_i$, where W_i is the width of level i, C_i is the maximum cost of any node at level i, and H is the height of the binary tree. Since the largest coefficient size of any of the polynomials at level i is dominated by $2^{2mi}d^{m+1}$ and the costs of all transformations are dominated by the cost of a translation, C_i is dominated by $m^3 i + m^3 L(d)$. Furthermore, $W_i \leq m$ for all i by Theorem 1, so the cost associated with level i is dominated by $m^4 i + m^4 L(d)$.

The height of the tree depends on the minimum separation (sep) between any two roots (real or complex) of the scaled polynomial $A(Bx)$. A node is bisected only if more than one variation is obtained. By Theorems 1 and 2, a node for an interval $[a, b]$ will be a leaf node if the circle centered at $(a+b)/2$ with diameter $2(b-a)$ contains at most one root, real or complex, and hence if $2(b-a) > \text{sep}(A)/B$. Since the diameter at level i is equal to 2^{-i+1}, this implies

$$\text{Height}(T) \leq \lceil \log_2(\text{sep}(A)^{-1}B) \rceil + 2.$$

By a theorem of Collins and Horowitz [4] or Mahler [16], the minimum root separation as a function of the degree and coefficient size is greater than $c_1(md)^{-c_2 m}$, where c_1 and c_2 are positive constants. This implies that $\log_2(\text{sep}(A)^{-1})$ is dominated by $mL(md)$. Therefore, H is dominated by $mL(md)$, and summing the costs of all of the levels, the computing time is dominated by $m^6 L(md)^2$. ∎

If each node at level i is computed in parallel, this bound on the maximum computing time can be improved to $m^5 L(md)^2$, since the width can be replaced with one. However, the computing time for the sequential algorithm can also be improved to $m^5 L(md)^2$ [11]. This result uses a new root separation theorem, due to Davenport [9]. Davenport's theorem implies that the width of higher levels must be small. Since Davenport's theorem does not improve the bound for the minimum root separation, it does not imply a better bound on the height of the tree. Therefore, it seems unlikely that parallelism combined with Davenport's result can be used to obtain a better parallel bound.

In Section 6.5, some data is presented which suggests that the average computing time is much smaller than either of these theorems predicts. Furthermore, it suggests that the average amount of parallelism, which corresponds to the average width, is small.

6.4 Parallel Implementation and Performance

The root isolation algorithm **IPRISMU** can be parallelized by performing the left and right tests and recursive calls in parallel. This parallel divide and conquer algorithm was implemented, on the Encore Multimax [10] (a shared memory multiprocessor), using PARSAC-2 [14], a parallel version of the SAC-2 computer algebra system [7]. This section describes our implementation and also serves as a non-trivial example of the use of PARSAC-2.

PARSAC-2 allows algorithms in the SAC-2 system to be executed in parallel using separate threads of control. A thread [8] is a lightweight process, which can execute a procedure with arguments. Since the fundamental data structure used by SAC-2 is a list, the threads used in PARSAC-2 must support parallel list processing. PARSAC-2 is based on S-threads [15]; these are threads that have access to a shared heap of list cells and are capable of performing, in parallel, the list operations required by SAC-2. Moreover, list cell allocation by S-threads is designed to minimize the contention due to multiple threads accessing the shared heap. Each S-thread has a local set of pages containing list cells. List cells are first allocated from the set of local pages, then if no cells are available locally, a new page is obtained from the shared heap. Upon completion of an S-thread, local pages can be transferred to other S-threads or returned to the heap for future use. For more details on the design of PARSAC-2 and the performance issues involved see [14].

The basic operations that can be performed on S-threads are *fork* and *join*. An S-thread can be forked to execute a procedure in parallel with the calling thread. Furthermore, an S-thread can later be joined to recover the result of its computation. S-threads are forked with "sthread_fork(func, arg, copyfunc)", where "func" is the name of the procedure to be executed, "arg" points to the arguments of "func", and "copyfunc" is a procedure for returning the result. If there are multiple arguments to "func", they must be packaged. "sthread_fork" returns a handle to the forked thread. S-threads are joined with "sthread_join(st)", where st is the handle of the thread to be joined.

S-threads allow three different mechanisms for passing SAC-2 list arguments: (1) by reference, (2) by transfer, and (3) by copy and transfer. In (1) only the pointer to a list is passed. This is possible in a shared memory implementation of S-threads. In (2) list arguments are transferred by transferring the set of pages containing them. Finally, in (3), the lists are copied to a fresh set of pages before transferring them. The parameter "copyfunc" to "sthread_fork" is used to copy the result of an algorithm when "sthread_join" is called. If "copyfunc" is NULL, then the result is transferred without being copied. After copying the result to a new set of pages, the old set of pages can be returned to the heap. Thus the intermediate garbage list cells are reclaimed. This procedure of returning list cells is called preventive garbage collection. This overview of S-threads contains enough information to understand the parallel implementation of **IPRISMU**. For more details concerning S-threads and S-thread operations see [15].

p_IPRISMU is a C implementation of the algorithm **IPRISMU**. It uses S-threads to support parallelism and algorithms from the SAC-2 library to perform the necessary polynomial operations. **p_IPRISMU** is a direct implementation of the algorithm listed in Section 6.2. The only difference is that it uses "sthread_fork" to compute the right test and recursive call in parallel with the left test and recursive call. **p_IPRISMU_r_shell** is used to compute the right half. A separate shell is needed to unpack the arguments. "sthread_join" is used to recover the intervals from the right half, which are then concatenated with the left and center intervals. Section 6.7 contains the listings of **p_IPRISMU** and **p_IPRISMU_r_shell**. Section 6.8 contains the specifications of the SAC-2 algorithms used by these programs.

p_IPRISMU was executed with random polynomials generated by the SAC-2 algorithm **IPRAN**. **IPRAN** generates r-variate integral polynomials with specified degree and coefficient sizes. The coefficients are uniformly distributed between -2^k and 2^k, where k is an input parameter. After a random polynomial is generated, it is scaled so that its roots lie in the unit circle.

Initially, polynomials of degrees 20 and 40 with coefficients less than 2^{10} in absolute value were used. The timings of **IPRISMU** and **p_IPRISMU** along with information concerning the search tree are reported in Table 6.1. The height of the tree H, the number of nodes N, and the average width, which is equal to N/H, are reported. All experiments were run on an Encore Multimax with 12 processors of type NS32332, rated at 2 MIPS each, and 64 MB main memory. To avoid influence of the system and other users in our timings, we limited the Encore Multimax to use only 8 of the 12 processors. Times were obtained from the Encore's microsecond wall-clock and are reported in milliseconds (ms).

Deg	Coef	IPRISMU	p_IPRISMU	H	N	Width	Speedup
20	10	463 ms	344 ms	5	6	1.200	1.346
20	10	631	337	5	8	1.600	1.873
20	10	350	157	3	5	1.667	2.229
20	10	279	170	3	4	1.333	1.643
20	10	354	169	3	5	1.667	2.097
40	10	1193	508	3	5	1.667	2.349
40	10	2929	1133	5	9	1.800	2.584
40	10	1208	492	3	5	1.667	2.457
40	10	2096	1104	6	8	1.667	1.899
40	10	4053	1616	7	12	1.714	2.509

TABLE 6.1. Parallel timings

The speedup is defined as the time for **IPRISMU** divided by the time for **p_IPRISMU**. For these experiments, a small amount of parallelism was observed. This is partly explained by the small average width of the trees. For each node, the left and right computation is done in parallel. Assuming the work associated with each node is the same, and is large compared to the overhead of S-threads, speedup should be approximately twice the average width. However, the cost of each node increases with its level in the tree and does not initially dominate the overhead cost. Also the trees tend to be thinner at the highest levels, because some roots have already been isolated at lower levels. Therefore the speedup observed in practice is somewhat smaller than that predicted by the average width. A more appropriate measure of parallelism would be a weighted average width which gives more weight to nodes at higher levels.

6.5 Average Behavior and Limits of Parallelism

In this section we investigate the average amount of parallelism obtainable with **p_IPRISMU**. Our study shows that the preliminary results obtained in Section 6.4 are typical. We ran our program for 1000 random polynomials of degrees 10 to 100 and computed the average height, width, and number of nodes of the corresponding search trees. From this data, we conjecture that on average there is limited parallelism available using the coefficient sign variation method. However, the efficiency of the algorithm is high and when more parallelism is available our program takes advantage of it. Polynomials with a large number of real roots are used to give an example where more parallelism is available. Finally, the data used for showing the limited amount of parallelism suggests that the average sequential computing time is significantly better than the worst case bound of Theorem 5.

Table 6.2 reports the average behavior of the coefficient sign variation method for the 1000 test cases. For each degree from 10 to 100, 100 random polynomials with coefficients of 10 bits were generated and various statistics of the algorithm were computed. For each polynomial, the height, average width, maximum width, and number of nodes of the corresponding search tree were computed. The number of real roots was also computed. Table 6.2 lists the average values of these measurements.

Degree	Real roots	Height	Nodes	Width	Max Width
10	2.14	3.60	5.10	1.35	1.68
20	2.58	4.05	5.93	1.45	1.86
30	2.80	4.07	6.28	1.54	1.93
40	2.92	4.45	6.83	1.53	1.96
50	3.14	4.69	7.08	1.53	1.98
60	3.10	4.37	6.64	1.54	2.00
70	3.05	4.20	6.58	1.58	2.00
80	3.28	5.03	7.68	1.54	2.00
90	3.38	5.01	7.66	1.55	2.03
100	3.74	5.25	7.98	1.54	2.03

TABLE 6.2. Average behavior

The average width appears to quickly approach a small constant value. This suggests a small constant bound for the average amount of parallelism.

Table 6.2 also indicates that the average height and number of nodes of the search trees are approximately proportional to the number of real roots. This relationship can be seen in the graph in Figure 6.4. This observation, if true in general, combined with a theorem of Kac's [12] on the average number of real roots, can be used to estimate the average computing time of **IPRISMU**. Kac's theorem states that the average number of real roots of a random polynomial is asymptotic to $\ln(m)$, where m is the degree of the polynomial. This theorem is true with several different definitions of random polynomials (see [2] for a more detailed discussion). In particular, it is true for our definition of random polynomials, i.e. those with uniformly distributed coefficients. This leads to the following conjecture, which is obtained by substituting $\ln(m)$ for the height H and a constant for the width W_i in the sum used in the proof of Theorem 5.

Conjecture 1 *Let* $\deg(A(x)) = m$. *Then the average computing time of the sign variation algorithm, when the coefficient size is fixed, is dominated by* $m^3\ln(m)^2$.

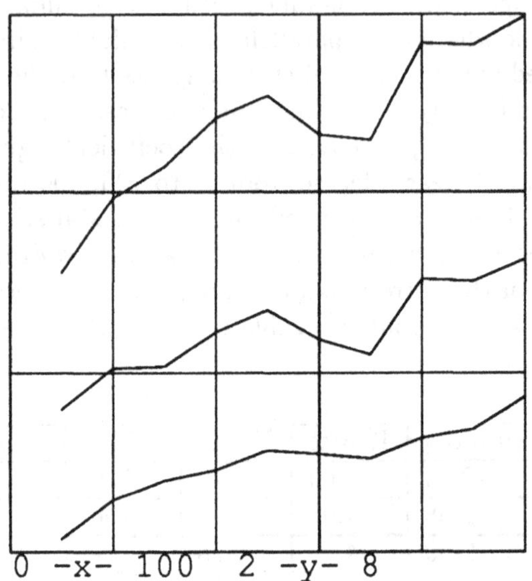

Figure 6.4. Number of real roots, height, and number of nodes as a function of degree.

The characteristics observed for random polynomials are not true for other classes of polynomials. In particular, if the number of real roots is increased then the width is increased. Therefore, for these polynomials, more parallelism is available. To see this, we generated polynomials with a specified number of real roots, uniformly distributed in the unit circle. For a given degree, $m = s + 2t$, a polynomial with s rational roots and t pairs of complex conjugate roots of the form $a \pm bi$, with a and b rational, were generated. Table 6.3 reports the behavior of **IPRISMU** and **p_IPRISMU** for polynomials of degree 20 and from 0 to 20 real roots. The graph in Figure 6.5 shows the relationship of speedup to the number of real roots.

6.6 Acknowledgements

This material is based upon work supported by the Ohio State University Office of Research and Graduate Studies under Award No. 221152, and by the National Science Foundation under Award Nos. CCR–9009396 and CCR–8913824.

6.7 Program Listings

```
#include "sthreads.h"
#include "sac2.h"

                         int  p_IPRISMU(A,a,b)
  int A,a,b;
/* Parallel version of IPRISMU.
*
*[Integral polynomial root isolation, symmetric coefficient
* sign variation method, unit circle, parallel algorithm.  A
* is a squarefree univariate integral polynomial.  a and b
* are binary rational numbers with a < b.  Let T be a
* rational linear fractional transformation which maps the
* interval [-1,1] to [a,b].  Let T(A) be an integral
* polynomial whose roots in [a,b] correspond to the roots of
* A in [-1,1].  Then the result L is a list (I_1,...,I_k)
* of one-point or open isolating intervals for the roots of
* T(A) in the interval (a,b).  I_1 < ... < I_k.]
*/

{ int L;  /* return variable. */
  int L_1, L_0, L_2;
  int A_,c,At,At2,v2,As,As2,At1,v1,As1,L1;
  int args[3];
  sthread_t th;
  extern int p_IPRISMU_r_shell ();

STEP1: /*[Initialize. Test right and right recursive call.]*/
  L_1 = NIL; L_0 = NIL; L_2 = NIL; L = NIL;
  A_ = PDBORD(A); c = RNSUM(a,b);  c = RNPROD(c,LIST2(1,2));
  args[0]=A_; args[1]=c; args[2]=b;
  th = sthread_fork(p_IPRISMU_r_shell,args,NULL);

STEP4: /*[Test midpoint.]*/
  if (PDEG(A_) != PDEG(A))  L_0 = COMP(LIST2(c,c),L_0);

STEP5: /*[Test left.]*/
  At = PRT(A_); At1 = IUPTR1(IUPNT(At));  v1 = IUPVAR(At1);
  if (v1 == 1)
    L_1 = COMP(LIST2(a,c),L_1);

STEP6: /*[Left recursive call.]*/
  if (v1 > 1)
    {As = IUPBHT(A_,-1);
```

```
     As1 = IUPNT(IUPTR1(IUPNT(As)));
     L_1 = p_IPRISMU(As1,a,c);
    }

STEP7:  /*[Concatenate list of intervals.]*/
   L_2 = sthread_join(th);
   L = CONC(L_0, L_2);
   L = CONC(L_1,L);
   return L;
}

                        int  p_IPRISMU_r_shell(arg)
int arg[3];
/* A shell for the right test and recursion of p_IPRISMU.
* The shell performs the right test on the polynomial arg[0].
* (arg[1], arg[2]) is the right half of the current interval.
* If more than one variation is found by the right test, the
* right transformation is done on arg[0] and p_IPRISMU is
* called.
*/

{int L_2;
 int k, P;
 int A_,c,b,At,At2,v2,As,As2;
 extern int p_IPRISMU ();

STEP1: /*[Initialize.]*/
  A_ = arg[0]; c = arg[1]; b = arg[2];

STEP2: /* Test right.*/
  At = PRT(A_);  At2 = IUPTR1(At);  v2 = IUPVAR(At2);
  if (v2 == 0)  L_2 = NIL;
  if (v2 == 1)  L_2 = COMP(LIST2(c,b),NIL);

STEP3: /*[Right recursive call.]*/
  if (v2 > 1)
    { As = IUPBHT(A_,-1); As2 = IUPTR1(As);
      L_2 = p_IPRISMU(As2,c,b);
    }

STEP4: /*[Transfer result.]*/
  return L_2;
}
```

6.8 SAC-2 Specifications

```
                    M:=COMP(a,L)
[Composition.  a is an object.  L is a list.  M is the
composition of a and L.]
                    L:=CONC(L1,L2)
[Concatenation.  L1 and L2 are lists.  L=CONC(L1,L2).  The
list L1 is  modified.]

                    A:=IPRAN(r,k,q,N)
[Integral polynomial, random.  k is a positive BETA-digit.
q is a rational number q1/q2 with 0 < q1 <= q2 < BETA.  N
is a list (n_r,...,n_1) of non-negative BETA-digits, r >= 0.
A is a random integral polynomial in r variables with
deg_i(A) <= n_i for 1 <= i <= r.  Max norm of A < 2^k and q
is the probability that any particular term of A has a
non-zero coefficient.]

                    B:=IUPBHT(A,k)
[Integral univariate polynomial binary homothetic trans-
formation.  A is a non-zero univariate integral polynomial.
k is a gamma-integer.  B(x)=2^{-h}A(2^{kx}) where h is
uniquely determined so that B is an integral polynomial not
divisible by 2.]

                    B:=IUPNT(A)
[Integral univariate polynomial negative transformation.  A
is a univariate integral polynomial.  B(x)=A(-x).]

                    b:=IUPRB(A)
[Integral univariate polynomial root bound.  A is an integral
polynomial of positive degree.  b is a binary rational number
which is a root bound for A.  If A(x) = \sum_{i=0}^n a_i x^i,
a_n /= 0, then b is the smallest power of 2 such that
2|a_{n-k}/a_n|^{1/k} <= b for 1 <= k <= n.  If a_{n-k}=0 for
1 <= k <= n then b=1.]

                    B:=IUPTR1(A)
[Integral univariate polynomial translation by 1.  A is a
univariate integral polynomial.  B(x)=A(x+1).]

                    n:=IUPVAR(A)
[Integral univariate polynomial variations.  A is a non-zero
univariate integral polynomial.  n is the number of sign
variations in the coefficients of A.]
```

```
                    B:=PDBORD(A)
[Polynomial divided by order.  A is a non-zero polynomial.
B(x) = A(x)/x^k where k is the order of A.]

                    n:=PDEG(A)
[Polynomial degree.  A is a polynomial.  n is the degree
of A.]

                    B:=PRT(A)
[Polynomial reciprocal transformation.  A is a non-zero
polynomial.  Let n=deg(A).  Then B(x) = x^nA(1/x), where x
is the main variable of A.]

                    T:=RNPROD(R,S)
[Rational number product.  R and S are rational numbers.
T=R*S.]

                    T:=RNSUM(R,S)
[Rational number sum.  R and S are rational numbers.
T=R+S.]
```

6.9 References

[1] A. Akritas and G. E. Collins. Polynomial real root isolation using Descartes' rule of signs. In *Proceedings of SYMSAC 76*, pages 272–275. ACM, New York, 1976.

[2] A. T. Bharucha-Reid. Random algebraic equations. In A. T. Bharucha-Reid, editor, *Probabilistic Methods in Applied Mathematics*, volume 2, pages 1–52. Academic Press, 1970.

[3] G. E. Collins. Computer algebra of polynomials and rational functions. *American Mathematical Monthly*, 80(7):725–755, August-September 1973.

[4] G. E. Collins and E. Horowitz. The minimum root separation of a polynomial. *Math. Comp.*, 28(126):589–597, April 1974.

[5] G. E. Collins and J. R. Johnson. Quantifier elimination and the sign variation method for real root isolation. In *Proceedings of the ACM-SIGSAM 1989 International Symposium on Symbolic and Algebraic Computation*, pages 264–271. ACM, 1989.

[6] G. E. Collins and R. Loos. Real zeros of polynomials. In B. Buchberger, G. E. Collins, and R. Loos, editors, *Computer Algebra*, pages 83–94. Springer-Verlag, Wien-New York, 1982.

[7] G. E. Collins and R. G. K. Loos. SAC-2 system documentation. Technical report. In Europe available from: R. G. K. Loos, Universität Tübingen, Informatik, D-7400 Tübingen, W-Germany. In the U.S. available from: G. E. Collins, Ohio State University, Computer Science, Columbus, OH 43210, U.S.A.

[8] E. C. Cooper and R. P. Draves. C threads. Technical report, Computer Science Department, Carnegie Mellon University, Pittsburg, PA 15213, July 1987.

[9] J. H. Davenport. Computer algebra for cylindrical algebraic decomposition. Technical report, The Royal Institute of Technology, Dept. of Numerical Analysis and Computing Science, S-100 44, Stockholm, Sweden, 1985.

[10] Encore Computer Corp. *Multimax Technical Summary*, 726-01759 Rev. E edition, January 1989.

[11] J. R. Johnson. *Algorithms for Polynomial Real Root Isolation.* PhD thesis, The Ohio State University, 1991. Dissertation in preparation.

[12] M. Kac. On the average number of real roots of a random algebraic equation. *Bulletin of the American Mathematical Society*, 49:314–320,938, 1943.

[13] D. E. Knuth. *The Art of Computer Programming (Seminumerical Algorithms)*, volume 2. Addison-Wesley, Reading, Mass., 1st edition, 1969.

[14] Wolfgang W. Küchlin. PARSAC-2: A parallel SAC-2 based on threads. In *Proc. AAECC-8*, LNCS, Tokyo, Japan, August 1990. Springer-Verlag. To appear.

[15] Wolfgang W. Küchlin. The S-threads environment for parallel symbolic computation. Chapter 1, these proceedings.

[16] K. Mahler. An inequality for the discriminant of a polynomial. *Michigan Mathematics Journal*, 11(3):257–262, September 1964.

Real roots	IPRISMU	p_IPRISMU	H	N	Width	speedup
0	2 ms	16 ms	1	1	1.000	0.107
2	440	224	2	3	1.500	1.961
4	432	220	2	3	1.500	1.964
6	955	524	4	6	1.500	1.822
8	1556	641	5	9	1.800	2.428
10	2154	656	5	12	2.400	3.284
12	2813	1048	7	15	2.143	2.686
14	3532	999	7	19	2.714	3.537
16	4632	1038	7	23	3.286	4.463
18	4780	1088	8	26	3.250	4.392
20	5086	1100	8	27	3.375	4.621

TABLE 6.3. Varying the number of real roots

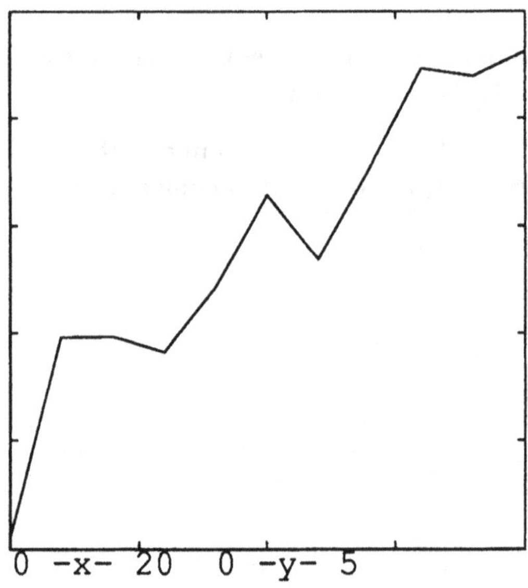

Figure 6.5. Speedup as a function of the number of real roots

7

Very Large Gröbner Basis Calculations

Winfried Neun[1]
Herbert Melenk[1]

Abstract: The attempt to solve systems of polynomial equations with Gröbner base techniques often leads to large problems which exceed the available computer resources with their requirements for cpu time or storage. The well-known reason for that is the swell of intermediate polynomials, which are generated during the basis calculation and which are in most cases not included in either the given set of polynomials or the resulting Gröbner basis. In this paper two different approaches to overcome the problem are presented which benefit from the usage of parallel computers, namely the vectorization of the arbitrary precision integer arithmetic and the usage of decomposition techniques. Especially the decomposition approach, where applicable, leads to massive parallelism in the problem solution, which results in a breakthrough for several problems.

7.1 Introduction

Within the group "Symbolik" at Konrad–Zuse–Zentrum Berlin (ZIB), the solution of systems of multivariate polynomial equations is one focus of research in the field of Computer Algebra. We try to find the solutions via the computation of Gröbner bases of the ideal generated by the systems polynomials. Therefore, we apply Buchberger's Algorithm [1] for the computation of a Gröbner basis on the system of polynomials. The solutions of the equation are then derived by postprocessing of the Gröbner basis polynomials. The computations are done on different types of computers with the Computer Algebra system REDUCE [5] especially using the (new) Gröbner package of Reduce [7] distributed via the REDUCE netlib by the Rand Corporation.

The computation of a Gröbner base is in many cases very expensive in terms of cpu time consumption and memory usage. It is not a rare case that a Gröbner base calculation takes several hours on a workstation system. Some Gröbner bases have been successfully computed within one week of operation. One of the well-known reasons for that behavior is that the intermediate polynomials consist of

[1]Konrad-Zuse-Zentrum für Informationstechnik Berlin, Heilbronner Str. 10, D-1000 Berlin 31, Federal Republic of Germany

many monomials whose coefficients are often extremely big, causing operations on such polynomials to be very slow. The polynomials in the given set of polynomials and in the resulting Gröbner bases are in most cases of moderate size. Just to give an impression of that effect we present the polynomial system given by GERDT and ZHARKOV [4] and the gcd of the coefficients of one of the intermediate polynomials.

Equations

$$\hat{e}_k = e_k = 0 \quad \text{for} \quad k = 1, \ldots 6$$

where $\hat{e}_k$ is derived from e_k 4 by exchanging a_i by b_i and

$$e_k = \hat{e}_k = 0, \quad (k = 1 + 6)\,,$$

where $\hat{e}_k$ is derived from e_k by exchanging a_i and b_i and

$$\begin{aligned}
e_1 &= a_1(a_3 - a_4) - a_4(b_3 - b_4)\,, \\
e_2 &= (2a_3 - a_4)y_1 - b_2 y_2\,, \quad y_1 = 6a_0 a_3 b_2 + (a_0 - b_0)(a_1^2 + a_4 b_2)\,, \\
e_3 &= a_2 y_1 - (2b_3 - b_4)y_2\,, \quad y_2 = 6a_0 a_2 b_3 + (a_0 - b_0)(a_1 a_2 + a_4 b_1)\,, \\
e_4 &= 3a_0(a_2 b_2 + a_3 b_3) + (a_0 - b_0)(a_1 + b_3)a_4\,,
\end{aligned}$$

$$\begin{aligned}
e_5 = {} & 2(2a_0^2 + 8a_0 b_0 - b_0^2)a_3 b_3 + 2(a_0 - b_0)(4a_0 - b_0)a_3 b_4 \\
& -6a_0(a_0 + 2b_0)a_2 b_2 + (a_0 - b_0)^2(5a_1 a_3 - 5a_1 a_4 + a_4 b_4) \\
& -(a_0 - b_0)7a_0 - b_0)a_4 b_3\,, \\
e_6 = {} & 3a_0(a_0 - b_0)[(a_0 - b_0)^3 - 3a_0(a_0 + 2b_0)^2](a_2 b_2 + a_3 b_3) \\
& +(a_0 - b_0)^3[3a_0 a_1 a_3 - 2(2a_0 + b_0)a_1 a_4] + 9a_0^2(a_0 - b_0)[(a_0 - b_0)a_4 \\
& -(a_0 + 2b_0)a_3]b_4 - (a_0 - b_0)(2a_0^3 - 30a_0^2 b_0 + b_0^3)a_4 b_3\,.
\end{aligned}$$

The GCD of the coefficients is:

```
3395051029924356607248256561587287203858529661768879171690104533450
1902795635721712167390074443168711597086445862171827118797255515357
0352927752921553348724899907144495652855901817009244253902634322509
7002604260065808335188052646714098380565761141056941088057900172916
226868589309070226629874986217937053128163965952000
```

Another problem related to the swell of expressions is the need for intermediate storage for the polynomials which in many cases exceeds the storage which can be allocated for a process.

Profiling the cpu time consumption of these computations shows that the basic arithmetic may cause up to 80 percent of the overall cpu time, while another great part is caused by memory management (garbage collection). All the other subtasks of the computation, e.g., the polynomial arithmetic, play a minor role in terms of cpu time consumption.

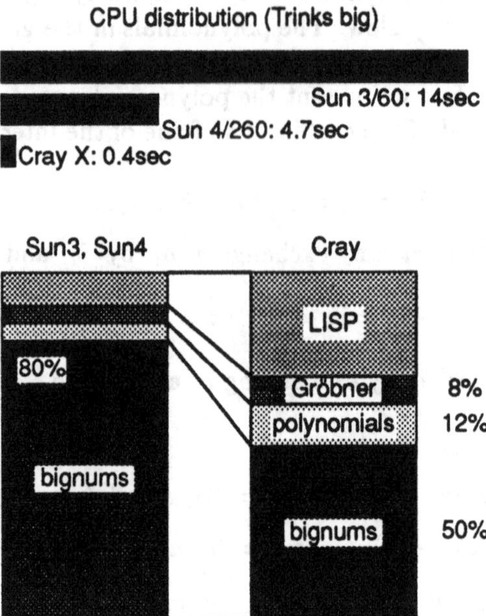

Figure 7.1.

Several attempts have been started to make these very large problems soluble in practice. The key idea is in any case to try to get as much cpu power (or as many cpu's) working on the problem as available. In this paper we will look at some aspects under development resp. done earlier in ZIB Berlin, which allow the usage of systems capable of parallel or vector operations.

Figure 7.1 shows the distribution of the cpu time consumption among the different tasks of the calculation on a workstation and on a Cray vector processor. It shows the effect of the vectorization, namely that the relative cpu time consumption of the bignum arithmetic is reduced drastically.

In the first part of this paper, we describe some possibilities for the usage of vector processing in the context of Gröbner base calculation and in the second part we look at some schemes for the reduction of problem strength by decomposing the polynomial system.

7.2 Bignum Package

A fast generic arithmetic system especially for the infinite precision integer arithmetic ("bignums") is a basic requirement for efficient computation of Gröbner bases. Therefore, we improved the generic arithmetic for the REDUCE systems used and distributed by ZIB, which includes vector processors like Cray machines and many workstation type computers.

For the workstations a conventional improvement is done. Since almost all of the micro processor cpus are equipped with instructions for double precision integer arithmetic, we use these instructions instead of the single precision version which is fully portable. This change together with some changes in the algorithms used, e.g. Lehmer gcd and Karatsuba type multiplication [6], we gained a significant speedup (see below).

In case of a vector processor, it is very natural to use the vector features for bignum arithmetic. The results of that approach implemented for Cray systems are discussed in [9], we will summarize the main points here. That implementation is of interest for other systems too, since it is not too Cray specific and vector processor features are not limited to supercomputers or mini–supercomputers any more. One can expect that the next generations of workstations will have vector capabilities too, e.g. Intel i860 or MIPS based workstation with optional vector processors are (as announced) already available.

At first, one has to find a way to generate vector instructions for specific Computer Algebra system. For the Cray version of REDUCE this was done by augmenting the compiler of the underlying LISP system (Portable Standard LISP). The vector interface defined here allows to code REDUCE or LISP procedures, which can access the vector hardware, e.g. vector registers and vector functional units like normal LISP storage resp. functions. The LISP systems uses the vector interface for other tasks than the generic arithmetic too, e.g. for garbage collection, which is described below. An external way to vector operation via FORTRAN or C was not chosen, since the foreign functional call is very expensive from the LISP system and should not be used on the basic operation level.

One of the main points for vectorizing an algorithm is the simplicity of basic operations and the uniformity of memory access. The vector length, i.e. the number of digits that can be handled in one operation, is important too, dependent on the machine type used, but in most cases not of prime interest. Many vector computers are equipped with vector registers which offer a very fast storage for the operands. Since the bandwidth between memory and the processing units is limited on most systems, a 'simple' algorithm which allows to keep the data in vector registers may be in many cases superior to a more sophisticated one without that property.

Bignums are implemented as densely packed structures in memory so there is a fast uniform data access. We handle the complete bignum or part of it by vector operations. If the vector hardware cannot be adapted to the actual operation e.g. the size of the operands is limited by the size of vector registers as on a Cray computer, we split the operation on the bignum in sufficiently small pieces.

The cpus of many modern computers consist of independent functional units which can work in parallel (e.g. multiply, add and logic units). Therefore, one should try to perform multiple operations on the stream of vector elements piped through the processor (e.g. adding and propagating the carry) simultaneously without storing intermediate results to memory.

The vectorized bignum arithmetic for REDUCE on Cray machines is using such principles of coding and achieves a significant linear speedup factor for relatively 'small' bignums already, without the need to pay for a startup penalty.

7.3 Results

In the following we give empirical results on the speed of bignum arithmetic implementations, namely the original and the improved ones for workstations and the vectorized version. The machines used here are Cray, SUN3 and SUN4. The tests are synthetic ones using two numbers of constant size. As one can see in the first two columns, the benefit gained by changing from single precision to double precision arithmetic instructions is great especially (as expected) for multiplication. The Cray architecture does not allow double precision multiplication in hardware, nor does the SUN4 (Sparc) architecture.

	SUN3 old bignum package (msec)	SUN3 new bignum package (msec)	SUN4 new bignum package (msec)	Cray vectorized bignums(msec)
add	1326	510	119	14
mult	20097	1660	952	7
div	6273	3111	1173	32

The next table gives the cpu times for the REDUCE standard test and the cpu time for the test of the (new) Gröbner package. The standard test, which does not use bignums, shows a ratio of 1/4 and 1/6 between Cray X–MP and the latest generation of workstations, whereas the Gröbner test suite uses bignums to a larger extent. Therefore, the ratio between Cray and workstation performance is bigger here, but the difference is not dramatic. We have to note here that the Gröbner package test suite does not contain extreme cases; it was designed to test functionality on many systems.

	REDUCE standard test (sec)	Gröbner package test suite (sec	normed: X-MP =1)
SUN3/60	20	74	2
SUN386i	20	74	21
Silicon Gr. Iris IP4	7	34	9
SUN4/SparcStation 1	5	20	6
DECstation 3100	4	18	5
SparcSystem	4	18	5
CD 4360	3	12.5	3 – 4
Cray-2	2	6.9	2
Cray X-MP	1	3.5	1
Cray Y-MP	0.8	2.9	0.8

It is worth noting that the block in the middle of the table above is built from workstations which are equipped with a Reduced Instruction Set (RISC) microprocessor. These new processors are very well suited for LISP applications since LISP does not use complicated form of memory access.

A detailed view on one example within the Gröbner test suite gives an impression of the speedup achievable with the vectorized bignum arithmetic. This example uses bignums to a larger extent than the other examples in the test suite.

Example : Trinks (big)

cpu time consumption: absolute and relative to Cray X-MP			
	Sun3/60	Sun4/60	Cray X-MP
sec:	14 (42)	4.7 (12)	0.4 (1)

7.4 Multi–Modular Bignum Representation

An alternative bignum representation of bignums using multiple moduli is presented in [9] too. It is based on the ability of a vector machine to process a certain amount of data in almost the same time as a single datum. Thus many modular computations can be done in roughly the same time as a single modular computation. Since there is no need for carry propagation, the multi modular representation allows a full exploitation of vector hardware speed, but the reconstruction using Chinese remainder algorithm is very expensive so that this method is valuable in special cases only.

7.5 Vectorized Garbage Collection

Though it is not directly connected to computation of Gröbner Bases, the problem of memory management and especially garbage collection is of great importance for the calculations. This is due to the fact that the intermediate polynomial operations allocate a lot of memory which is unused soon afterwards. In very large calculations, the garbage collector overhead will be responsible for up to 60 percent of the overall computation time. Therefore, the is a big potential for optimization in the choice of an adequate garbage collection technique.

In Gröbner basis calculations one observes a special behavior of the memory space allocation. Typically, after a startup phase where many polynomials are generated which are contained in the result, for a very long time only temporary polynomials are produced, which will vanish soon afterwards. That means that a large amount of stable information must be kept while a garbage collection is needed very often.

The garbage collector used on the Cray systems is optimized towards this behavior. A compacting variant is used because of the limitations of real memory. Normally, after a certain time of computation, a great amount of memory is occupied and relatively stable.

One key idea is to reduce the number of walks over the stable part (the 'old' polynomials) and to compact in the new space (the 'new' polynomials) only. The stable part is inspected at each garbage collection to keep track of destructive changes in the old structures, but it is compacted only from time to time or when a lack of memory occurs.

The next step the storage of pairs and other LISP items is separated such that the storage for pairs can be inspected and compacted using vector techniques. The same strategy for the old storage is used twice now. In this way, the biggest parts of the static storage can be treated by vector hardware such that the overall garbage collector time can be reduced by a factor of up to 5, depending on the application.

7.6 Small Grain Parallelism

In several papers (e.g.[8], [10]), it has been pointed out that Buchberger's Algorithm for the computation of a Gröbner basis of an ideal cannot be implemented easily for a parallel system using a great number of processors in an efficient way. This is a bit surprising since the Buchberger Algorithm looks as though it allows parallel operation while working on a huge set of pairs. The main reason for that can be found in the "criteria", which allow to predict whether or not one can cancel an H-polynomial computation because the result will be zero. In many cases, the number of nonzero H-polynomials is much lesser than the number of predicted zero H-polynomials. So one cannot abandon the application of the criteria without doing lots of extra work. This imposes that a parallel variant of Buchberger's Algorithm needs a lot of communication and control between subtasks, which makes an implementation complicated and reduces the benefit one can achieve by using parallel hardware.

7.7 Decomposition of Polynomial Systems

An alternative approach to the ones described above is the decomposition of the problem into several subproblems which can be solved independently. For large problems this leads to massive parallel operations, which may be run on every configuration of parallel hardware because it does not depend on the fast synchronization between subtasks. It can be run effectively on a network of workstations by socket communication or with a simple fork/join mechanism on a multiprocessor system.

7.8 Factorization

If an H-polynomial generated during the Gröbner base calculation is factorable, the complexity of the problem can be reduced in many cases. In this case one can split the problem of calculation of a Gröbner base in several subproblems, which can be run in parallel then. Since we are not interested in the original Gröbner Base as such, but in the solutions of a system of equations, we can split the problem and calculate partial solutions of the problems derived from Gröbner bases of super ideals of the original one. The union of the partial solutions contains the solution of the original problem.

Since the factorization in general leads to smaller systems, there is a good chance that one can solve the partial problems while one cannot (reasonably) solve the original one. The partial problems can be run completely independent from each other without the need for further synchronization, one has to take care about multiple computations of the same subproblem.

In practical computations it turned out that factorization often leads to massive parallelism, and the problem of keeping track of the factorization process to avoid repeated calculations of the same partial solution is not trivial. In the example from GERDT/ZHARKOV [4], the calculation splits the original problem into 212 subproblems.

Although the way of splitting the systems under work via factorization of an H–polynomial is simple and very effective, there are some limitations for that approach.

The test for factorability is relatively expensive and for big problems many polynomials will be tested without success. Buchberger's Algorithm defines a sequence of H–polynomials which will be computed during the calculation of the basis. Although the ideal may contain factorable polynomials, it will happen very often that they are not seen during the calculation, since they are not in the set of intermediate polynomials.

Because the naive idea to wait at the stream of intermediate polynomials is not satisfactory in many cases because none of the polynomials is factorable, we try to construct factorable polynomials in the ideal generated by the intermediate polynomials ourselves.

7.9 Active Factorization

This approach is a heuristic one. Assuming we find a factoring intermediate polynomial during the calculation, one can try to find the same factors in the ideal earlier already. In some other cases the user already knows some partial solutions of his problem, because he understands the background on which the equations are built.

In these cases one may try to find factorable polynomials using the assumed factors as target, i.e. given a potential factor T and a set of H–polynomials H we construct the sets of quotients of H and T and the sets of remainders of H and T. If we find a nontrivial representation of zero in the ideal of the remainders, we build the polynomial Q with the same representation among the quotients. Then $T * Q$ is a factorable polynomial in the original ideal.

7.10 Splitting Polynomial Systems in Case of Symmetry

Many polynomial systems are 'symmetric', which means that one can interchange or cyclically rotate some variable without changing the polynomial system. A class of symmetric systems given by H. CAPRASSE/ J. DEMARET [2] is the following where one can change the usage of all variables.

$$3x_1^3 - x_1 3(x_1 a_1 - a_1 - a_2 + 2) + a_2 - a_3 = 0,$$
$$3x_2^3 - x_2 3(x_2 a_1 - a_1 - a_2 + 2) + a_2 - a_3 = 0,$$
$$3x_3^3 - x_3 3(x_3 a_1 - a_1 - a_2 + 2) + a_2 - a_3 = 0,$$
$$3x_4^3 - x_4 3(x_4 a_1 - a_1 - a_2 + 2) + a_2 - a_3 = 0,$$
$$3x_5^3 - x_5 3(x_5 a_1 - a_1 - a_2 + 2) + a_2 - a_3 = 0,$$

$$a_i := x_1^i + x_2^i + x_3^i + x_4^i + x_5^i \,, \quad i = 1, 2, 3 \,.$$

These systems of polynomials turned out to be soluble (with high cost) up to the order of five and with testfactors supplied by H. CAPRASSE. But the problems of higher order were completely insoluble with reasonable effort on existing hardware with standard Gröbner base algorithms. This is caused by the fact that the standard algorithm does not benefit from the symmetric nature of the original problem, moreover, new polynomials generated by Buchberger's algorithm are not symmetric any more and the systems looses its symmetric property.

The factorization of the systems polynomials and new intermediate polynomials normally fails or succeeds only after hours of run time. For several types of symmetric problems, including the one above, K. GATERMANN [3] recently described and implemented algorithms which either split the polynomial a priori in several smaller systems or the systems is transformed such that the polynomials will factorize immediately. As mentioned above, factoring the system polynomials or in general splitting the system allows to use parallel operation within a very simple fork and join scheme.

The systems above are soluble now up to the order of $n = 9$, which still takes a week or two (in sequential mode) to be solved on a fast workstation. For $n = 9$, the number of subproblems which are generated during the a priori decomposition is limited by $256(2^{n-1})$, so there is a lot of work that can be done in parallel with a reasonable amount of processors, which turns out to be a breakthrough at least for several types of problems.

For $n = 5$ the following polynomials are added automatically to generate a subproblem:

$$\begin{aligned} &x_1^2 - x_1x_2 - x_1x_3 - x_1x_4 - x_1x_5 + x_1 + x_2^2 - x_2x_5 + x_2 \\ &\quad + x_3^2 - x_3x_5 + x_3 + x_4^2 + x_4x_5 + x_4 + x_5^2 + x_5 - 2, \\ &a_2 - x_1a_1 + a_1 - x_3(x_2 + x_4 + x_5) - 2, \\ &x_1 - x_4, \\ &x_1 - x_2 \end{aligned}$$

The partial solution belonging to the subproblem is obtained in a few seconds now:

$$\{x_1, x_2, x_3 - 1, x_4, x_5\}$$

$$\{1281x_1 - 1560x_5^5 + 20067x_5^4 - 55916x_5^3 + 12773x_5^2 + 10350x_5 + 2346,$$

$$1281x_2 - 1560x_5^5 + 20067x_5^4 - 55916x_5^3 + 12773x_5^2 + 10350x_5 + 2346,$$

$$1281x_3 - 1560x_5^5 + 20067x_5^4 - 55916x_5^3 + 12773x_5^2 + 10350x_5 + 2346,$$

$$x_4 - x_5,$$

$$13x_5^6 - 176x_5^5 + 580x_5^4 - 432x_5^3 - 12x_5^2 + 36x_5 + 18\}$$

$$\{x_1, x_2, x_3, x_4, x_5 - 1\}$$

7.11 Conclusions

We attack the problem of computation of complicated Gröbner bases in two different ways, namely by improving the underlying LISP system of the computer algebra system REDUCE especially for vector processors like Cray X-MP and by application of decomposition methods. By vectorizing the bignum arithmetic we obtain speedups up to a factor of 100 versus the latest generation of workstation computers. The algorithms implemented for the vector operations with bignums can be easily migrated to different vector architectures. The second attempt leads to systems which can be easily run on loosely coupled systems or on multiprocessor systems. The impact of a decomposition for the solution of a system of equations can hardly be over estimated. The combination of both seems to be boding very well for the next generation of super workstations which are equipped with more than one processor and/or with vector processing features too.

7.12 REFERENCES

[1] B. Buchberger: *Gröbner bases: An algorithmic method in polynomial ideal theory.* In N. K. Bose (ed.): Multidimensional Systems Theory, pp. 184–232, D. Reidel Publ. Comp (1985).

[2] H. Caprasse, J. Demaret: *private communication.*

[3] K. Gatermann: *Symbolic solution of polynomial equation systems with symmetry.* Preprint SC 90-3, Konrad-Zuse-Zentrum Berlin (1990). Accepted for ISSAC 1990.

[4] V. P. Gerdt, A. Yu Zharkov: *Computer Classification of Integrable Coupled KdV-like Systems.* Preprint E5–89–232, Joint Institute for Nuclear Research, Dubna (1989).

[5] A. C. Hearn: *REDUCE User's Manual, Version 3.3.* The Rand Corporation, Santa Monica (1987).

[6] D. E. Knuth: *The Art of Computer Programming: Seminumerical Algorithms.* Vol. **2**, 2nd edition (1981).

[7] H. Melenk, H. M. Moeller, W. Neun: *GROEBNER: A Package for Calculating Gröbner Bases.* REDUCE NetLib, The Rand Corporation (1990).

[8] H. Melenk, W. Neun: *Parallel polynomial operators in the large Buchberger algorithm.* In J. Della Dora, J. Fitch (eds): Computer Algebra and Parallelism, Academic Press (1988).

[9] W. Neun, H. Melenk: *Implementation of the LISP- arbitrary precision arithmetic for a vector processor.* In J. Della Dora, J. Fitch (eds): Computer Algebra and Parallelism, Academic Press (1988).

[10] C. Ponder: *Evaluation of "Performance Enhancements" in Algebraic Manipulation Systems.* In J. Della Dora, J. Fitch (eds): Computer Algebra and Parallelism, Academic Press (1988).

8

Boolean Gröbner Bases and Their MIMD Implementation

Pascale Senechaud[1]

Abstract: We present two methods to compute Gröbner bases in parallel, both based on Buchberger's sequential algorithm. A distributed memory MIMD computer (the FPS 140) gives experimental results obtained with boolean polynomials.
The algorithms were implemented on the FPS T40 connected as a ring and as a hypercube of processors. The first implementation shows the interest of the parallelization. The second one, based on a divide and conquer strategy, has a behavior very close to the sequential algorithm.
We evaluate the contribution of the parallelism by a direct comparison of sequential and parallel times without references to complexity.

8.1 Introduction

Some parallel studies of the Buchberger algorithm can be found in the literature: a vectorization of the basic arithmetic operations [5] (Cray-2 XMP) and two parallelizations on shared memory multiprocessors [9] (16 processor Encore Machine) and [6] (Alliant, 4 processors). Here, the point of view is different. Indeed, all the computers used in these different cases are shared memory and the problems encountered are not the same, in particular because of the memory access, which is very important in the Buchberger algorithm.

The machine we use has p processors (each with a transputer, a vectorboard unit and a local memory of 1Mbyte) connected according to a hypercube topology. Communications are based on a "rendez-vous" protocol between neighbor processors.

Let us recall some basic notions on Gröbner bases [8]. Let K be a field and $A = K[x_1, \ldots, x_n]$ be a ring of polynomials in $x_1, \ldots, x_n$ over K. Let I be an ideal of A given by a finite set of polynomials $\{f_1, \ldots, f_p\}$ (Hilbert). Let $<$ be a well defined total order on the monomials with coefficient 1 (power product). For every polynomial p, we define its initial part Init(p): if $m(p)$ is the leading power product and $c(p)$ the corresponding coefficient then $\text{Init}(p) = c(p)m(p)$.

We use the notion of a division between multivariate polynomials that had been

[1]Faculté des Sciences, Dept. de mathematique, 123 Av. A. Thomas, 87060 Limoges Cedex, France

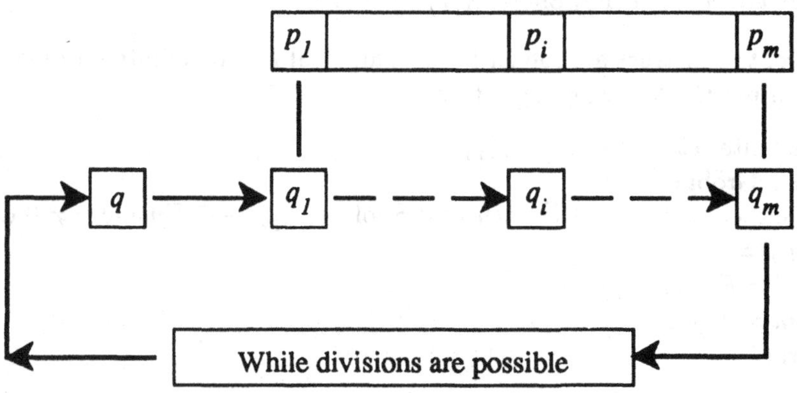

Figure 8.1. Multiple Reduction

introduced by Hironaka in [1]: let p $(= \text{Init}(p) + m_2 + \cdots + m_j)$ and q $(= \text{Init}(q) + n_2 + \cdots + n_k)$ be two polynomials of A. The "division" of p by q is defined by $r = q - a \times p$ where

$$a = \begin{cases} \text{monomial if it exists } n_i = a\,\text{Init}(p) \\ 0 \text{ if every } i \text{ if } n_i \neq a\,\text{Init}(p) \end{cases}$$

It is easy to extend this operation to the division (also named *reduction*) of any family $F = \{p_1, \ldots, p_m\}$ by a polynomial q (see Figure 8.1).

We denote by $-^*F$ the reflexive transitive closure of this reduction relation associated to F. One can show that $-^*F$ is noetherian and computes *the normal form modulo* F of a polynomial q ($-^*F$ is the *normalization of a polynomial*).

Let us introduce the following notations:

$\alpha = (\alpha_1, \ldots, \alpha_n)$ the exponent of Init(p), $c(p)$ the corresponding coefficient,

$\beta = (\beta_1, \ldots, \beta_n)$ the exponent of Init(q), $c(q))$ the corresponding coefficient,

$\delta = (\delta_1, \ldots, \delta_n)$ such that $\forall i, \delta_i = \max(\alpha_i, \beta_i)$,

Spoly$(p, q) = c(q)X^{\delta-\alpha}p - c(p)X^{\delta-\beta}q$ (X is the power product on all the variables and Spoly(p, q) is the *spolynomial of* p *and* q).

A set of polynomials F is a *Gröbner basis* of the ideal I if for all $p \in A$ $p \in I \Longleftrightarrow p \longrightarrow^* F0$. The construction of the Gröbner basis is allowed by the following theorem [2]:

Theorem 6 (Buchberger) *Let I be an ideal of A generated by $F = \{p_1, \ldots, p_m\}$. The following properties are equivalent:*

- *F is a Gröbner basis*

- *For all* $(p_i, p_j) \in F^2$, $\text{Spoly}(p_i, p_j) \longrightarrow^* F0$.

Therefore, to construct a Gröbner basis of an ideal I from a finite set of generators, F, we can use the following algorithm:

Input: a finite set $F = \{p_1, \ldots, p_m\}$.
Output: a Gröbner basis.
While there are $(i, j) \in I^2$ such that $\text{Spoly}(p_i, p_j) \longrightarrow^* Fp$ and $p \neq 0$ do
$p_m = p$
$F = F \cup \{p_{m+1}\}$
$m = m + 1$
return(F)

The Gröbner basis obtained is not unique. However, from an arbitrary basis, we can obtain the reduced Gröbner basis which is unique [4]. This particular basis is obtained by adding to the former algorithm the **reduction** of each p_i to its normal form modulo p_j (for all $j \neq i$) (*the reduction of a polynomial set*).

8.2 The Parallel Algorithms

The time T_n to transfer n contiguous bytes of memory from a processor to its neighbors is $T_n = \beta + n\tau$ where β is the communication link set up time and τ the time to transfer one byte. For the FPS T40, $\beta/\tau = 61.73$, therefore, we prefer to transfer sets of polynomials that lie in contiguous areas of memory rather individual polynomials.

We must insist on the data sharing and on the way chosen to number of the input polynomials: the execution time depends on these in a very acute manner. We have chosen an equidistributed sharing. Obviously, the distribution of m polynomials on p processors can be done in several manners. In order to compare the computing times for different p (number of processors used), we choose the same sharing strategy for each p.

8.2.1 Using a Ring

Let F be the input family. Let m be the number of polynomials to be distributed among p processors, such that $m = pq + r$. After the distribution, each processor has in its local memory k polynomials, where $k = q$ for $p - r$ processors, and $k = q + 1$ for the remaining r processors. The k polynomials of the ith processor are called the reference polynomials and are denoted by FP_i.

The basic idea of the algorithm on a a ring of processors, is the distribution of the independent tasks of the algorithm between the processors. The parallel algorithm is build from *calculus modules* (BCM) and *principles of communication* (PC).

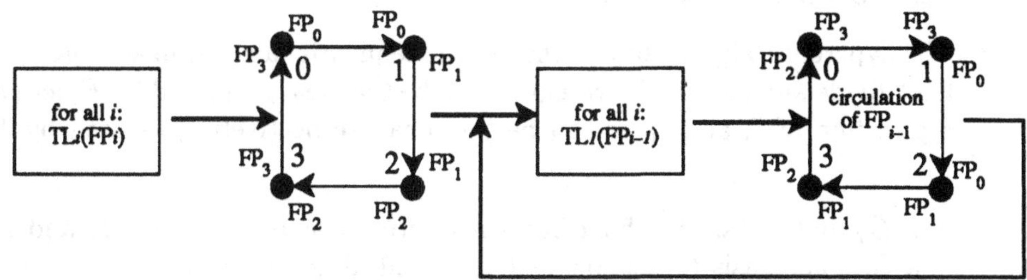

Figure 8.2. First stage of parallel algorithm

Stage 1:

- The first BCM is

  $TL_i(FP) ==$
  input: a finite set of FP polynomials.
  output: SPN_i; spolynomials associated to pairs formed with a polynomial of FP and a polynomial of FP_i and then normalized modulo FP_i.

- The first PC is a movement of the reference polynomials along a ring.

- If we compose the BCM given above and the PC we obtain the first stage of the parallel algorithm as shown in Figure 8.2. The polynomials (called SPN_i) associated to F and locally normalized modulo P_i are obtained after $p/2$ executions of the elementary sequence [BCM, PC].

Stage 2: The normalization of the SPN_i uses the same scheme as before. The BCM is then a stage of normalization (NormFF(FP, FP_i)) where the polynomial set FP is normalized modulo FP_i. The same PC is employed with the SPN_i. These two operations are composed, updating FP (equal to FP_i, then equal to $FP_{i-1}, \ldots$). In order to finish this composition and obtain the SPN_i normalized, modulo all the reference polynomials, we must generate the synchronous termination of this computation. This is done by means of message exchanges.

We note MW the composition of (i) and (ii), corresponding to one iteration of the main "while loop" of the sequential algorithm presented in the introduction. For each processor MW has FP_i as input and gives the SPN_i (normalized) as output. At each iteration of the inputs to MW become $FP_i = FP_i \cup SPN_i$. The loop is stopped when the normal form of all the SPN_i are equal to zero, in all the processors. A sending of messages controls the synchronous termination of the computation.

In order to obtain the Gröbner basis, we introduce the reduction. It is inserted, after each modification of the inputs to MW, without parallelization because it is intrinsically sequential.

8.2.2 USING A HYPERCUBE

For the hypercube algorithm we consider the sequential algorithm as a basic module with a "divide and conquer" strategy. Let I be the ideal generated by F; let (F_1, F_2) be a partition of I, and I_1 and I_2 be the ideals respectively generated by F_1 and F_2:

- If G_1 and G_2 are Gröbner bases respectively associated with I_1 and I_2 then a Gröbner basis G_{12} associated to the ideal generated by $G_1 \cup G_2$ is also a Gröbner basis associated by I.
- If G_1 and G_2 are reduced, G_{12} is the Gröbner basis G associated with I.

We use the fact that G_1 and G_2 are already Gröbner bases and we adapt the Buchberger algorithm in this particular case (see [7] for details). Let Buch2 be the corresponding procedure.

We apply the principle given above on four processors. Let F be the input family. We proceed in the same way in the ring and divide F into local families. Let PE_1, PE_2, PE_3 and PE_4 be the four processors. In the first stage each PE_i computes the Gröbner basis associated with its input obtaining four sub-bases G_1, G_2, G_3 and G_4. Through communication procedures, we distributes these sub-bases in order to have G_1 and G_2 in the processor PE_1, G_3 and G_4 in the processor PE_4. Applying Buch2, we can obtain in these two processors respectively two new sub-bases G_{12} and G_{34}. Then PE_1 sends G_{12} to PE_4, which computes the complete Gröbner basis associated with F (using G_{12} and G_{34}).

During the call of Buch2 to compute G_{12} and G_{34} processors PE_2 and PE_3 are idle. In order to use them we construct another data distribution: after communication, processor PE_2 contains G_3 and G_1 and processor PE_4 contains G_2 and G_4. These two processors call Buch2 and compute G_{31} and G_{24} respectively. Then processor PE_2 sends G_{31} to processor PE_3 which computes the Gröbner basis associated with F.

At each stage, the execution time depends on the data sharing, the the two ways to compute the complete Gröbner basis require the same length of time. The algorithm is stopped as soon as we obtain the Gröbner basis in a processor. Let Hyp4 be the procedure which, from the local families, computes the Gröbner basis in the way we have just described.

We generalize this construction to 2^d processors ($d \geq 3$). We assign a task to a cube of dimension d. We divide the two subtasks assigned to two sub-cubes of dimension $d-1$ putting together (by pairs) the data of the cube (shared between the different processors of the cube) in two different ways. Let $REP(d)$ be the communication procedure that gives these two ways. Each processor uses Buch2 to compute the Gröbner basis associated to its pair. We obtain, in that way 2^{d-2} cubes with four processors using Hyp4 giving us the Gröbner basis in $2^d/2$ different ways. The algorithm is stopped on all processors as soon as the Gröbner basis is obtained on one processor.

This is illustrated in Figure 8.3. The procedure is as follows

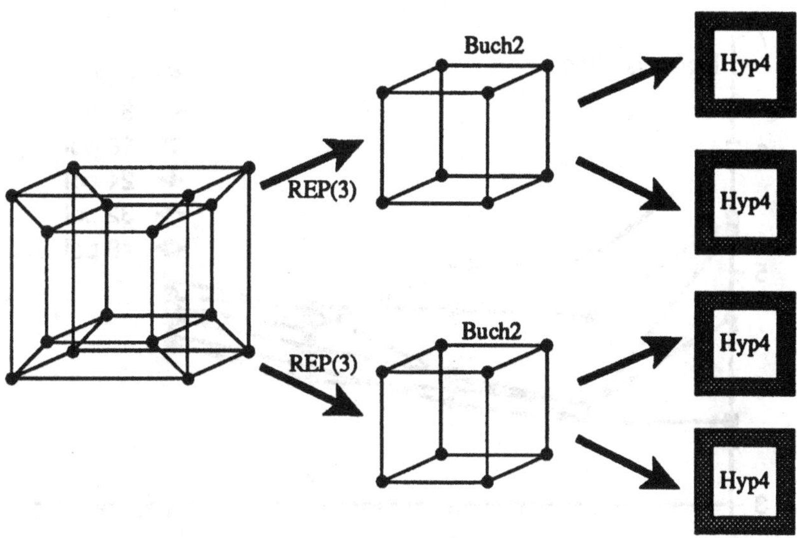

Figure 8.3. Hypercube Gröbner Basis Computation with 16 Processors

1. After sharing the polynomials, the first stages corresponds to the computation of 16 sub-bases (one for each processor).

2. The second stage is the reunion of the former sub-bases into two repartitions of height pairs. Using $REP(3)$ we obtain two repartitions: (the basis contained in the ith processor is denoted by I.

   A (1, 2); (3, 4); (5, 6); (7, 8); (9, 10); (11, 12); (13, 14); (15, 16).

   B (1, 16); (2, 15); (3, 14); (4, 13); (5, 12); (6, 11); (7, 10); (8, 9).

3. On both independent sub-cubes with eight processors:
   - we compute the eight corresponding sub-bases using Buch2.
   - We call $REP(2)$ to obtain two repartitions of four pairs of sub-bases.

4. A call to Hyp4 is performed on each cube containing four processors.

8.3 Experimental Results

8.3.1 Implementation on a Ring of Processors

The most efficient initial reduction of the polynomials consists in reducing all the polynomials by each other before dispatching them on the ring. The result of this reduction is not known *a priori* which makes difficult the generation of a wide variety of input polynomial sets. Thus, we build some input sets (assumed to be

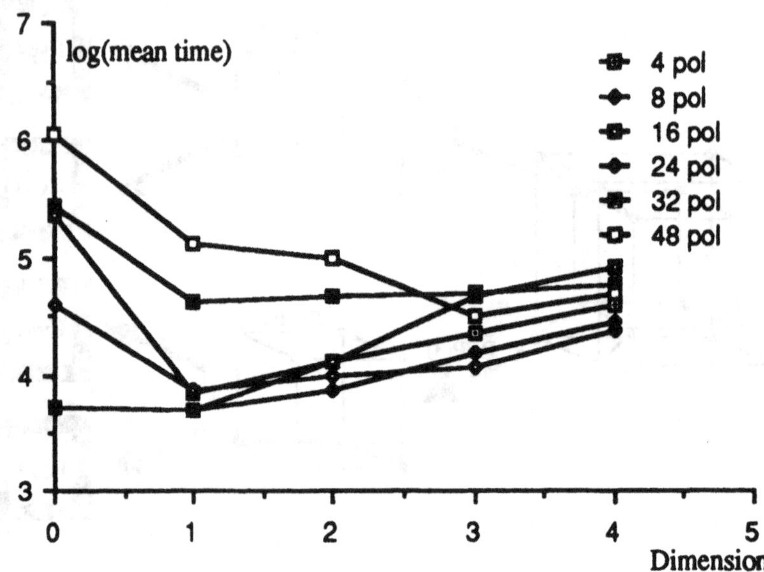

Figure 8.4. Log of Mean Execution Time

the result of the reduction of unknown bigger sets) and distribute them on the ring. The initial reduction becomes a preprocessing of the input data.

An interpretation is possible—in spite of the fluctuations of the execution time—if we limit the studies to homogeneous examples: we choose polynomials of 1 to 5 monomials in 15 indeterminates. The mean times introduced are henceforth obtained by computing the mean of the execution times of some examples (from 7 to 10 depending on the cases) with the same number of polynomials. Let $\mathcal{A}(d)$ be the ring of dimension d (with 2^d processors). We represent the decimal logarithm of the mean time to execute a test with q polynomials, according to d in Figure 8.4.

When we treat a small number of polynomials, the rings with few processors are the best performing. ON the other hand, if th enumber of polynomials is large enough, the rings with many processors perform better. $\mathcal{A}(0)$ treats examples with 4 and 8 polynomials in the same times. $\mathcal{A}(1)$ treats with equivalent times examples with 4, 8 16, 24 polynomials and to treat examples with 32 polynomials it takes 10 times longer. $\mathcal{A}(3)$ has the same behavior but the time to treat examples with 32 polynomials is only 2.5 times longer.

Thus we define the *capacity* of a ring of a given dimension to treat problems of variable size: it is the number of initial polynomials per processor from which we note a significant variation of the execution time for a given dimension (here nearly 3 or 4).

We call **grading** the numbering of the rings of different size in function of increasing efficiency. To exploit the results more quantitatively we represent the mean of the grading of the processors according to the number of polynomials (see

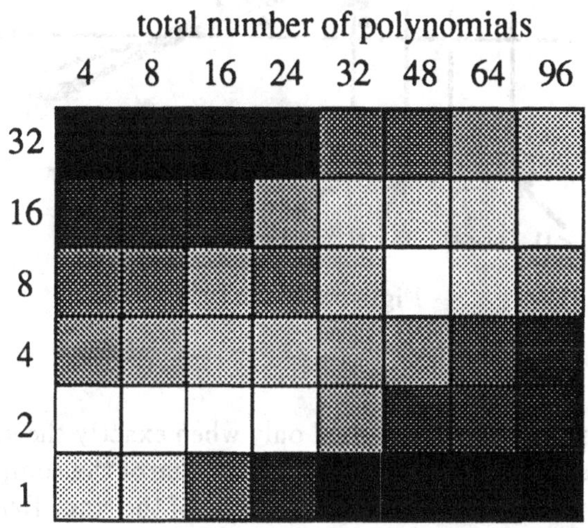

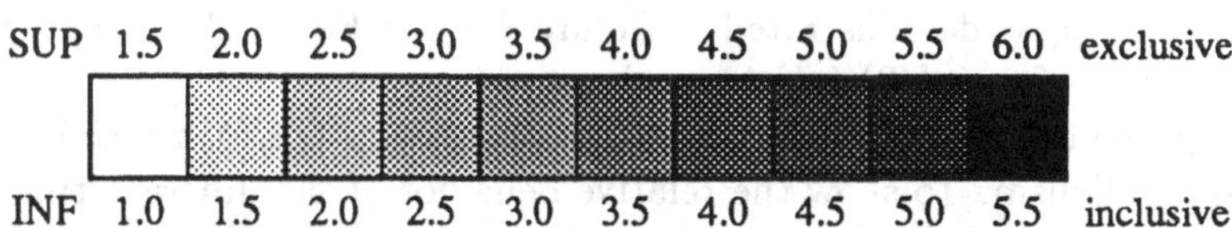

Figure 8.5. Grading of polynomials

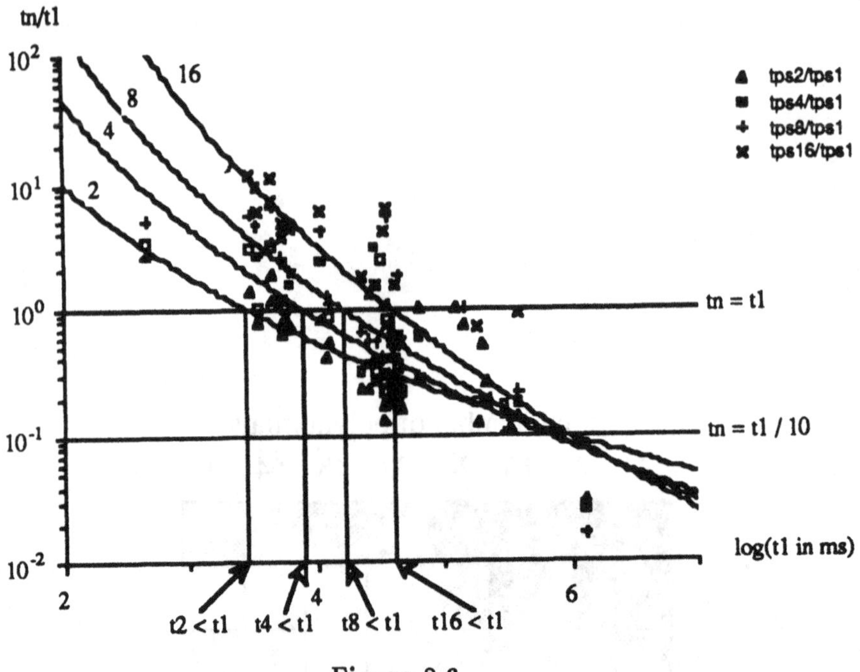

Figure 8.6.

Figure 8.5).

Generally, the concept of efficiency is used only when exactly the same problem is treated with one processor or with n, i.e., when the value of the computations made during the algorithm is independent of the dimension of the ring. Here the variation of this volume is too important to use the notion of efficiency in a classical manner (in reality we do not execute the same algorithm when we change the number of processors). Thus, as demonstrated in Figure 8.6 and Figure 8.7, we represent the ratio of the time of are $\mathcal{A}(0)$ (t1) according to t1.

These representations show us the range of processor rings for which it is interesting to use parallelism. To show the relative behavior of the different rings, we give a zoom (Figure 8.7) of the first diagram (Figure 8.6) when the decimal logarithm of t1 varies from 5 to 7.

8.3.2 Implementation on a Hypercube of Processors

The initial reduction is an integral part of the computation algorithm: the input data are automatically reduced. The examples used in the ring case lead to very short times when executed on a hypercube (ratios of about 100) and thus we decided to generate larger initial sets, which were already reduced, to obtain higher execution times. We treat examples with:

- from 10 to 96 polynomials
- between 2 and 10 monomials per processor

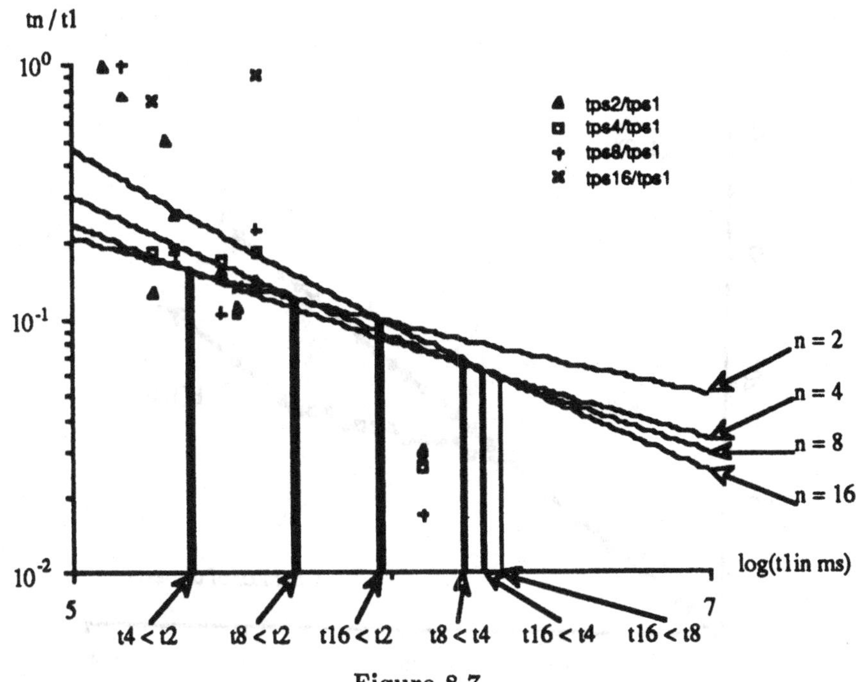

Figure 8.7.

- from 2 to 20 indeterminates

We denote by $\mathcal{H}(d)$ a hypercube of dimension d.

If we represent the execution times obtained with n processors according to the time obtained with one processor, we notice the same global behaviors as in the ring case. For short execution times we have the grading of processors, 1, 2, 4, 8, 16 whence for great execution times we have the reverse grading. Nevertheless, this behavior is obviously smaller with $\mathcal{H}(d)$ than with $\mathcal{A}(d)$. The difference between $\mathcal{H}(0)$ and $\mathcal{H}(1)$is not so pronounced as between $\mathcal{A}(0)$ and $\mathcal{A}(1)$ [7].

To show the behavior of $\mathcal{H}(d)$ we propose the following study: We consider N tests E_i computed on $\mathcal{H}(d)$ and $\mathcal{H}(d')$ and ordered in increasing $T_i(d)$ (execution time on $\mathcal{H}(d)$). We define the boolean variable $\mathrm{Sup}_j(d', d)$ with $d' > d$ and the function $F_i(d', d)$ as follows:

$$\mathrm{Sup}_j(d', d) = \begin{cases} 1 & \text{if } T_j(d') > T_j(d) \\ 0 & \text{if } T_j(d') < T_j(d) \end{cases}$$

$$F_i(d', d) = \sum_{j=1}^{i} \mathrm{Sup}_j(d', d)$$

If we represent $F_i(D', d)$ and i according to i the two curves should be equal for small values of i and should become different as i increases. The first one must become asymptotically horizontal. If $\mathcal{H}(d')$ is always worse than $\mathcal{H}(d)$ after a certain value of i, the quantity $F_i(d', d) - i$ stays at the same value. The following diagrams represent F with (d', d) equal to $(1, 0)$, $(2, 0)$, $(3, 0)$ and $(2, 1)$ successively.

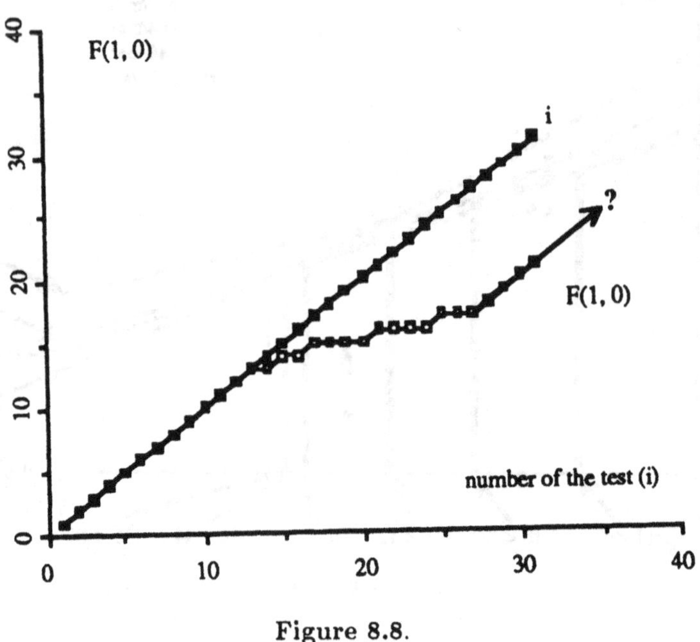

Figure 8.8.

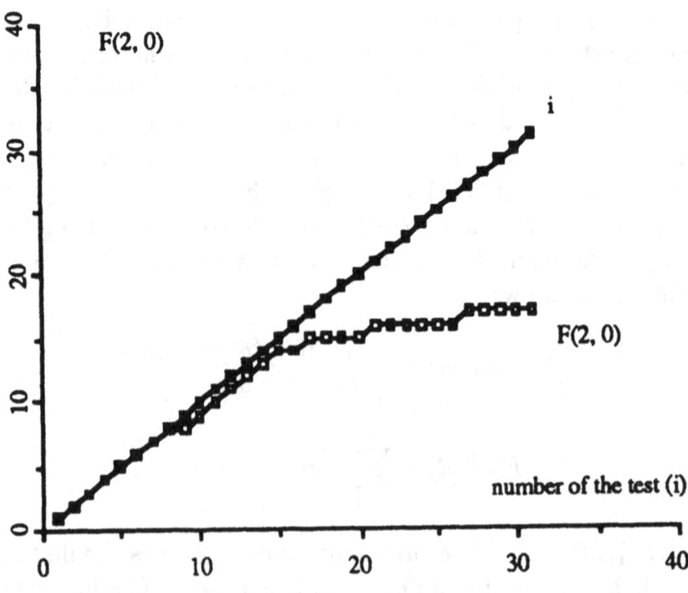

Figure 8.9.

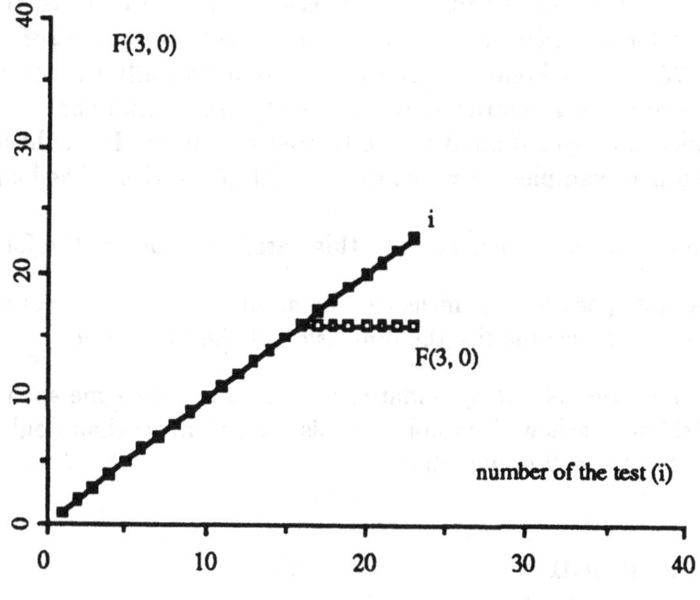

Figure 8.10.

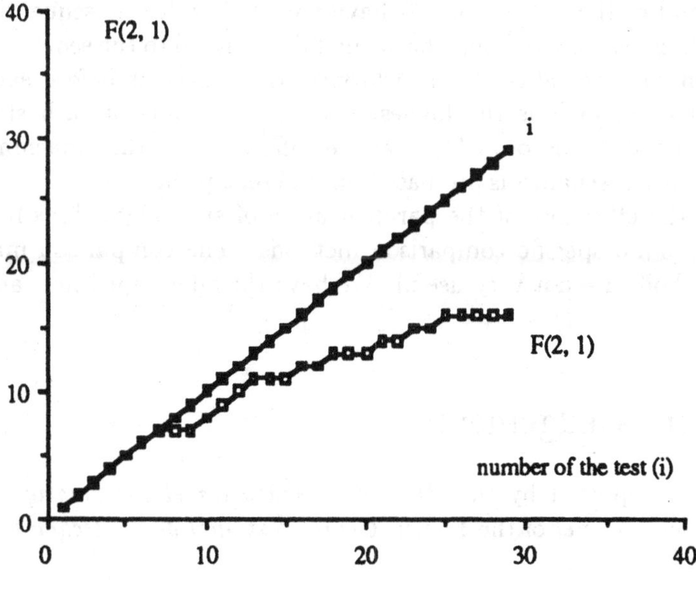

Figure 8.11.

Of these different diagrams we can see that as long as $T_i(d)$ is small, $F_i(d', d)$ is equal to i and after a fixed value of $T_i(d)$ is smaller than i. Thus, the parallelism is interesting for hypercubes containing more than two processors. Indeed the behavior of $\mathcal{H}(1)$ is different. It seems that asymptotically the algorithm on two processors becomes systematically worse than the sequential one.

These conclusions are limited to the treated examples. It would be interesting to examine other examples for which the sequential algorithm has longer execution times.

The main problems we encounted in this parallelization are the following:

- We cannot speak about efficiency: the number of arithmetic operations is not constant when we modify the dimension of the hypercube.
- If the behavior is not systematic, it is because the time spent to compute the Gröbner basis with m polynomials is much more than double of the time associated to $m/2$ polynomials.

8.4 Conclusion

The quantitative study of the first parallel algorithm—where the independent tasks are executed in parallel and where the polynomials run along a ring—shows the interest of the parallelization in particular for the data which have an important number of polynomials. The second algorithm based on a "divide and conquer" strategy—where the number of communications depends only on the size of the hypercube and of the data—has a behavior very close to the sequential algorithm because of its construction from a basic module identical to the sequential algorithm (executed on smaller data). We synchronize the processor before each procedure of communication so it is the highest computation time of each stage that the total execution depends on. This can be different on other machines when the communications mechanisms are based on different principles.

To study the efficiency of the parallelization of such algorithms it seems to be necessary to build specific comparison methods. The comparison made with the classical methods are not very useful. We have thus developed new analysis methods.

8.5 Acknowledgements

This work is supported by the *PRC Mathématiques et Informatique* and by the *GRECO Calcul Formel* of the French *Centre National de la Recherche Scientifique (CNRS).*

8.6 References

[1] J. M. Aroca, H. Hironaka, J. L. Vincente, "The theory of Maximal Contact," *Memorias de Mathematica del Instituto "Jorge Juana,"* 29, (1975).

[2] B. Buchberger, "Ein algorithmisches Kriterium für die Lösbarkeit eines algebraischen Gleichungsystems," *Aequationes Math*, 4, pp. 374–383, (1970).

[3] B. Buchberger, "A criterion for detecting unnecessary reductions in the construction of Gröbner bases," *Proc. EUROSAM 79 (Ed. W. Ng)*, Springer LNCS 72, pp. 3–21, (1979).

[4] B. Buchberger, "Gröbner Bases: an algorithmic method in polynomial ideal theory," *Recent Trends in Multidimensional Systems Theory (Ed. N. K. Bose)*, Reidel, (1985).

[5] H. Melenk, H. M. Möller, W. Neun, "On Gröbner Basis Computation on a Supercomputer Using Reduce," FB Mathematik und Informatik der Fernuniversität Hagen, Preprint SC 88–2, (1988).

[6] C. Ponder, "Parallelism Algorithms for Gröbner Basis Reduction," Electrical Engineering and Computer Sciences Dept, University of California, Berkeley, CA, (1988).

[7] P. Sénéchaud, "Bases de Gröbner Booléennes Méthodes de Calcul; Applications; Parallelisation," Thèse INPG, (1990).

[8] W. Trinks, "On B. Buchberger's Method for Solving Algebraic Equations," *J. Number Theory*, 10/4, pp. 475–488, (1978).

[9] J. P. Vidal, "The Computation of Gröbner Bases on a shared memory multiprocessor," Technical Report, School of Computer Science, Carnegie Mellon University, to appear.

[10] S. Watt *Bounded Parallelism in Computer Algebra*, Ph. D. Thesis, University of Waterloo, Waterloo, Ontario (1985).

Lecture Notes in Computer Science

For information about Vols. 1–504
please contact your bookseller or Springer-Verlag

Vol. 505: E. H. L. Aarts, J. van Leeuwen, M. Rem (Eds.), PARLE '91. Parallel Architectures and Languages Europe, Volume I. Proceedings, 1991. XV, 423 pages. 1991.

Vol. 506: E. H. L. Aarts, J. van Leeuwen, M. Rem (Eds.), PARLE '91. Parallel Architectures and Languages Europe, Volume II. Proceedings, 1991. XV, 489 pages. 1991.

Vol. 507: N. A. Sherwani, E. de Doncker, J. A. Kapenga (Eds.), Computing in the 90's. Proceedings, 1989. XIII, 441 pages. 1991.

Vol. 508: S. Sakata (Ed.), Applied Algebra, Algebraic Algorithms and Error-Correcting Codes. Proceedings, 1990. IX, 390 pages. 1991.

Vol. 509: A. Endres, H. Weber (Eds.), Software Development Environments and CASE Technology. Proceedings, 1991. VIII, 286 pages. 1991.

Vol. 510: J. Leach Albert, B. Monien, M. Rodríguez (Eds.), Automata, Languages and Programming. Proceedings, 1991. XII, 763 pages. 1991.

Vol. 511: A. C. F. Colchester, D.J. Hawkes (Eds.), Information Processing in Medical Imaging. Proceedings, 1991. XI, 512

Vol. 512: P. America (Ed.), ECOOP '91. European Conference on Object-Oriented Programming. Proceedings, 1991. X, 396 pages. 1991.

Vol. 513: N. M. Mattos, An Approach to Knowledge Base Management. IX, 247 pages. 1991. (Subseries LNAI).

Vol. 514: G. Cohen, P. Charpin (Eds.), EUROCODE '90. Proceedings, 1990. XI, 392 pages. 1991.

Vol. 515: J. P. Martins, M. Reinfrank (Eds.), Truth Maintenance Systems. Proceedings, 1990. VII, 177 pages. 1991. (Subseries LNAI).

Vol. 516: S. Kaplan, M. Okada (Eds.), Conditional and Typed Rewriting Systems. Proceedings, 1990. IX, 461 pages. 1991.

Vol. 517: K. Nökel, Temporally Distributed Symptoms in Technical Diagnosis. IX, 164 pages. 1991. (Subseries LNAI).

Vol. 518: J. G. Williams, Instantiation Theory. VIII, 133 pages. 1991. (Subseries LNAI).

Vol. 519: F. Dehne, J.-R. Sack, N. Santoro (Eds.), Algorithms and Data Structures. Proceedings, 1991. X, 496 pages. 1991.

Vol. 520: A. Tarlecki (Ed.), Mathematical Foundations of Computer Science 1991. Proceedings, 1991. XI, 435 pages. 1991.

Vol. 521: B. Bouchon-Meunier, R. R. Yager, L. A. Zadek (Eds.), Uncertainty in Knowledge-Bases. Proceedings, 1990. X, 609 pages. 1991.

Vol. 522: J. Hertzberg (Ed.), European Workshop on Planning. Proceedings, 1991. VII, 121 pages. 1991. (Subseries LNAI).

Vol. 523: J. Hughes (Ed.), Functional Programming Languages and Computer Architecture. Proceedings, 1991. VIII, 666 pages. 1991.

Vol. 524: G. Rozenberg (Ed.), Advances in Petri Nets 1991. VIII, 572 pages. 1991.
pages. 1991.

Vol. 525: O. Günther, H.-J. Schek (Eds.), Advances in Spatial Databases. Proceedings, 1991. XI, 471 pages. 1991.

Vol. 526: T. Ito, A. R. Meyer (Eds.), Theoretical Aspects of Computer Software. Proceedings, 1991. X, 772 pages. 1991.

Vol. 527: J.C.M. Baeten, J. F. Groote (Eds.), CONCUR '91. Proceedings, 1991. VIII, 541 pages. 1991.

Vol. 528: J. Maluszynski, M. Wirsing (Eds.), Programming Language Implementation and Logic Programming. Proceedings, 1991. XI, 433 pages. 1991.

Vol. 529: L. Budach (Ed.), Fundamentals of Computation Theory. Proceedings, 1991. XII, 426 pages. 1991.

Vol. 530: D. H. Pitt, P.-L. Curien, S. Abramsky, A. M. Pitts, A. Poigné, D. E. Rydeheard (Eds.), Category Theory and Computer Science. Proceedings, 1991. VII, 301 pages. 1991.

Vol. 531: E. M. Clarke, R. P. Kurshan (Eds.), Computer-Aided Verification. Proceedings, 1990. XIII, 372 pages. 1991.

Vol. 532: H. Ehrig, H.-J. Kreowski, G. Rozenberg (Eds.), Graph Grammars and Their Application to Computer Science. Proceedings, 1990. X, 703 pages. 1991.

Vol. 533: E. Börger, H. Kleine Büning, M. M. Richter, W. Schönfeld (Eds.), Computer Science Logic. Proceedings, 1990. VIII, 399 pages. 1991.

Vol. 534: H. Ehrig, K. P. Jantke, F. Orejas, H. Reichel (Eds.), Recent Trends in Data Type Specification. Proceedings, 1990. VIII, 379 pages. 1991.

Vol. 535: P. Jorrand, J. Kelemen (Eds.), Fundamentals of Artificial Intelligence Research. Proceedings, 1991. VIII, 255 pages. 1991. (Subseries LNAI).

Vol. 536: J. E. Tomayko, Software Engineering Education. Proceedings, 1991. VIII, 296 pages. 1991.

Vol. 537: A. J. Menezes, S. A. Vanstone (Eds.), Advances in Cryptology – CRYPTO '90. Proceedings. XIII, 644 pages. 1991.

Vol. 538: M. Kojima, N. Megiddo, T. Noma, A. Yoshise, A Unified Approach to Interior Point Algorithms for Linear Complementarity Problems. VIII, 108 pages. 1991.

Vol. 539: H. F. Mattson, T. Mora, T. R. N. Rao (Eds.), Applied Algebra, Algebraic Algorithms and Error-Correcting Codes. Proceedings, 1991. XI, 489 pages. 1991.

Vol. 540: A. Prieto (Ed.), Artificial Neural Networks. Proceedings, 1991. XIII, 476 pages. 1991.

Vol. 541: P. Barahona, L. Moniz Pereira, A. Porto (Eds.), EPIA '91. Proceedings, 1991. VIII, 292 pages. 1991. (Subseries LNAI).

Vol. 542: Z. W. Ras, M. Zemankova (Eds.), Methodologies for Intelligent Systems. Proceedings, 1991. X, 644 pages. 1991. (Subseries LNAI).

Vol. 543: J. Dix, K. P. Jantke, P. H. Schmitt (Eds.), Nonmonotonic and Inductive Logic. Proceedings, 1990. X, 243 pages. 1991. (Subseries LNAI).

Vol. 544: M. Broy, M. Wirsing (Eds.), Methods of Programming. XII, 268 pages. 1991.

Vol. 545: H. Alblas, B. Melichar (Eds.), Attribute Grammars , Applications and Systems. Proceedings, 1991. IX, 513 pages. 1991.

Vol. 546: O. Herzog, C.-R. Rollinger (Eds.), Text Understanding in LILOG. XI, 738 pages. 1991. (Subseries LNAI).

Vol. 547: D. W. Davies (Ed.), Advances in Cryptology – EUROCRYPT '91. Proceedings, 1991. XII, 556 pages. 1991.

Vol. 548: R. Kruse, P. Siegel (Eds.), Symbolic and Quantitative Approaches to Uncertainty. Proceedings, 1991. XI, 362 pages. 1991.

Vol. 549: E. Ardizzone, S. Gaglio, F. Sorbello (Eds.), Trends in Artificial Intelligence. Proceedings, 1991. XIV, 479 pages. 1991. (Subseries LNAI).

Vol. 550: A. van Lamsweerde, A. Fugetta (Eds.), ESEC '91. Proceedings, 1991. XII, 515 pages. 1991.

Vol. 551:S. Prehn, W. J. Toetenel (Eds.), VDM '91. Formal Software Development Methods. Volume 1. Proceedings, 1991. XIII, 699 pages. 1991.

Vol. 552: S. Prehn, W. J. Toetenel (Eds.), VDM '91. Formal Software Development Methods. Volume 2. Proceedings, 1991. XIV, 430 pages. 1991.

Vol. 553: H. Bieri, H. Noltemeier (Eds.), Computational Geometry - Methods, Algorithms and Applications '91. Proceedings, 1991. VIII, 320 pages. 1991.

Vol. 554: G. Grahne, The Problem of Incomplete Information in Relational Databases. VIII, 156 pages. 1991.

Vol. 555: H. Maurer (Ed.), New Results and New Trends in Computer Science. Proceedings, 1991. VIII, 403 pages. 1991.

Vol. 556: J.-M. Jacquet, Conclog: A Methodological Approach to Concurrent Logic Programming. XII, 781 pages. 1991.

Vol. 557: W. L. Hsu, R. C. T. Lee (Eds.), ISA '91 Algorithms. Proceedings, 1991. X, 396 pages. 1991.

Vol. 558: J. Hooman, Specification and Compositional Verification of Real-Time Systems. VIII, 235 pages. 1991.

Vol. 559: G. Butler, Fundamental Algorithms for Permutation Groups. XII, 238 pages. 1991.

Vol. 560: S. Biswas, K. V. Nori (Eds.), Foundations of Software Technology and Theoretical Computer Science. Proceedings, 1991. X, 420 pages. 1991.

Vol. 561: C. Ding, G. Xiao, W. Shan, The Stability Theory of Stream Ciphers. IX, 187 pages. 1991.

Vol. 562: R. Breu, Algebraic Specification Techniques in Object Oriented Programming Environments. XI, 228 pages. 1991.

Vol. 563: A. Karshmer, J. Nehmer (Eds.), Operating Systems of the 90s and Beyond. Proceedings, 1991. X, 285 pages. 1991.

Vol. 564: I. Herman, The Use of Projective Geometry in Computer Graphics. VIII, 146 pages. 1992.

Vol. 565: J. D. Becker, I. Eisele, F. W. Mündemann (Eds.), Parallelism, Learning, Evolution. Proceedings, 1989. VIII, 525 pages. 1991. (Subseries LNAI).

Vol. 566: C. Delobel, M. Kifer, Y. Masunaga (Eds.), Deductive and Object-Oriented Databases. Proceedings, 1991. XV, 581 pages. 1991.

Vol. 567: H. Boley, M. M. Richter (Eds.), Processing Declarative Kowledge. Proceedings, 1991. XII, 427 pages. 1991. (Subseries LNAI).

Vol. 568: H.-J. Bürckert, A Resolution Principle for a Logic with Restricted Quantifiers. X, 116 pages. 1991. (Subseries LNAI).

Vol. 569: A. Beaumont, G. Gupta (Eds.), Parallel Execution of Logic Programs. Proceedings, 1991. VII, 195 pages. 1991.

Vol. 570: R. Berghammer, G. Schmidt (Eds.), Graph-Theoretic Concepts in Computer Science. Proceedings, 1991. VIII, 253 pages. 1992.

Vol. 571: J. Vytopil (Ed.), Formal Techniques in Real-Time and Fault-Tolerant Systems. Proceedings, 1992. IX, 620 pages. 1991.

Vol. 572: K. U. Schulz (Ed.), Word Equations and Related Topics. Proceedings, 1990. VII, 256 pages. 1992.

Vol. 573: G. Cohen, S. N. Litsyn, A. Lobstein, G. Zémor (Eds.), Algebraic Coding. Proceedings, 1991. X, 158 pages. 1992.

Vol. 574: J. P. Banâtre, D. Le Métayer (Eds.), Research Directions in High-Level Parallel Programming Languages. Proceedings, 1991. VIII, 387 pages. 1992.

Vol. 575: K. G. Larsen, A. Skou (Eds.), Computer Aided Verification. Proceedings, 1991. X, 487 pages. 1992.

Vol. 576: J. Feigenbaum (Ed.), Advances in Cryptology - CRYPTO '91. Proceedings. X, 485 pages. 1992.

Vol. 577: A. Finkel, M. Jantzen (Eds.), STACS 92. Proceedings, 1992. XIV, 621 pages. 1992.

Vol. 578: Th. Beth, M. Frisch, G. J. Simmons (Eds.), Public-Key Cryptography: State of the Art and Future Directions. XI, 97 pages. 1992.

Vol. 579: S. Toueg, P. G. Spirakis, L. Kirousis (Eds.), Distributed Algorithms. Proceedings, 1991. X, 319 pages. 1992.

Vol. 580: A. Pirotte, C. Delobel, G. Gottlob (Eds.), Advances in Database Technology – EDBT '92. Proceedings. XII, 551 pages. 1992.

Vol. 581: J.-C. Raoult (Ed.), CAAP '92. Proceedings. VIII, 361 pages. 1992.

Vol. 582: B. Krieg-Brückner (Ed.), ESOP '92. Proceedings. VIII, 491 pages. 1992.

Vol. 583: I. Simon (Ed.), LATIN '92. Proceedings. IX, 545 pages. 1992.

Vol. 584: R. E. Zippel (Ed.), Computer Algebra and Parallelism. Proceedings, 1990. IX, 114 pages. 1992.